Analog Circuits and Signal Processing

Series Editors:

Mohammed Ismail, Dublin, USA
Mohamad Sawan, Montreal, Canada

The Analog Circuits and Signal Processing book series, formerly known as the Kluwer International Series in Engineering and Computer Science, is a high level academic and professional series publishing research on the design and applications of analog integrated circuits and signal processing circuits and systems. Typically per year we publish between 5–15 research monographs, professional books, handbooks, edited volumes and textbooks with worldwide distribution to engineers, researchers, educators, and libraries.

The book series promotes and expedites the dissemination of new research results and tutorial views in the analog field. There is an exciting and large volume of research activity in the field worldwide. Researchers are striving to bridge the gap between classical analog work and recent advances in very large scale integration (VLSI) technologies with improved analog capabilities. Analog VLSI has been recognized as a major technology for future information processing. Analog work is showing signs of dramatic changes with emphasis on interdisciplinary research efforts combining device/circuit/technology issues. Consequently, new design concepts, strategies and design tools are being unveiled.

Topics of interest include:
Analog Interface Circuits and Systems;
Data converters;
Active-RC, switched-capacitor and continuous-time integrated filters;
Mixed analog/digital VLSI;
Simulation and modeling, mixed-mode simulation;
Analog nonlinear and computational circuits and signal processing;
Analog Artificial Neural Networks/Artificial Intelligence;
Current-mode Signal Processing;
Computer-Aided Design (CAD) tools;
Analog Design in emerging technologies (Scalable CMOS, BiCMOS, GaAs, heterojunction and floating gate technologies, etc.);
Analog Design for Test;
Integrated sensors and actuators;
Analog Design Automation/Knowledge-based Systems;
Analog VLSI cell libraries;
Analog product development;
RF Front ends, Wireless communications and Microwave Circuits;
Analog behavioral modeling, Analog HDL.

More information about this series at http://www.springer.com/series/7381

Kerim Türe • Catherine Dehollain
Franco Maloberti

Wireless Power Transfer and Data Communication for Intracranial Neural Recording Applications

 Springer

Kerim Türe
École Polytechnique Fédérale de Lausanne
Lausanne, Switzerland

Catherine Dehollain
École Polytechnique Fédérale de Lausanne
Lausanne, Switzerland

Franco Maloberti
University of Pavia
Pavia, Italy

ISSN 1872-082X   ISSN 2197-1854   (electronic)
Analog Circuits and Signal Processing
ISBN 978-3-030-40828-2   ISBN 978-3-030-40826-8   (eBook)
https://doi.org/10.1007/978-3-030-40826-8

This Springer imprint is published by the registered company Springer Nature Switzerland AG.
The registered company address is: Gewerbestrasse 11, 6330 Cham, Switzerland

To My Father, My Mother,
My Brother and My Sister,

and

To My Beloved Wife

– Kerim Türe

Preface

The impact of technological advancement became significant and still has an essential role to play in the diagnosis and treatment of diseases. Nowadays, personalized medicine and therapy are on the rise as a consequence of the developed and cheaper medical technologies which enables the monitoring and evaluation of many parameters of patients. One of the areas where the technological developments in biomedical applications can be used most effectively is the neurological monitoring systems. The brain–machine interface (BMI) allows controlling external software or hardware by transmitting neural information. Besides, BMI can be used for recovering lost functions such as walking. In addition, as brain signals can be analyzed in a more efficient way, advanced monitoring can be used to understand the cause and to treat neurological disorders such as epilepsy or Parkinson's diseases. The development of closed-loop systems is still going on to detect the abnormalities in neural signals and prevent them by applying short electrical pulses.

A neurological disorder is a chronic discomfort that affects many people throughout the world. The major form of treatment is long-term drug therapy, and surgery is an alternative for patients who do not respond to drug treatment when a seizure is limited to one area. Continuous monitoring of neural activity is commonly used for the diagnosis of epileptic zones. Thanks to the advances in electronics, sensor systems, and fabrication methodologies; this book proposes an entirely implantable wireless neural monitoring system which eliminates the wires and possible complications due to them.

The power requirement of the active components in the implanted system is supplied by a hybrid solution composed of wireless power transmission and a rechargeable battery. A 4-coil inductive link is used for remote powering. A stable and ripple-free supply voltage for the implanted system is generated by the use of an active half-wave rectifier and low drop-out regulator blocks. Additionally, the automatic resonance tuning mechanism is developed to maximize power transfer efficiency. Furthermore, the power feedback generation structure is designed to make sure that the delivered power is reached.

Wireless data communication solutions between the external unit and the implanted system are presented. Configuration parameters of the implanted system are transferred by modulating the powering signal source and the embedded electronics decode changes in the powering signal. Two alternative methods, namely narrowband and ultra-wideband transmitters, are designed to transmit the amplified, quantized, and analyzed neural activities. The transmitters can be used interchangeably to provide longer transmission distance or higher data rates.

Finally, the designed electronic circuits are fabricated and packaged with biocompatible materials for in vivo characterization. The operation of remote powering and wireless data transmission units are validated in a rat.

Lausanne, Switzerland
Lausanne, Switzerland
Pavia, Italy
December 2019

Kerim Türe
Catherine Dehollain
Franco Maloberti

Contents

Acronyms

ADC	Analog-to-Digital Converter
ASK	Amplitude Shift Keying
BER	Bit Error Rate
BMI	Brain–Machine Interface
BPF	Band-Pass Filter
BPSK	Binary Phase Shift Keying
CC-LNA	Capacitive-Coupled Low-Noise Amplifier
CE	Conformité Européenne
CFN	Capacitive Feedback Network
CMOS	Complementary Metal-Oxide-Semiconductor
CSF	Cerebrospinal Fluid
DFF	D Flip-Flop
DSP	Digital Signal Processing
ECoG	Electrocorticography
EEG	Electroencephalogram
EIRP	Effective Isotropic Radiated Power
EM	Electromagnetic
EP	Endocochlear Potential
EU	European Union
FCC	Federal Communication Commission
FDA	Food and Drug Administration
FF	Far-Field
FoM	Figure of Merit
FSK	Frequency Shift Keying
HFO	High-Frequency Oscillations
iEEG	Intracranial Electroencephalogram
IR-UWB	Impulse Radio Ultra-Wideband
ISM	Industrial, Scienctific and Medical
ISO	International Organization for Standardization
LDO	Low Drop-Out
LE	Less than Enough

LFP	Local Field Potentials
LNA	Low-Noise Amplifier
LO	Local Oscillator
Mbps	Megabits per Second
ME	More than Enough
MEA	Multielectrode Array
mi	Modulation Index
Mpps	Megapulse per second
MSB	Most Significant Bit
NEF	Noise Efficiency Factor
NF	Near-Field
OOK	On-Off Keying
OPAMP	Operational Amplifier
OTA	Operational Transconductance Amplifier
PA	Power Amplifier
PCB	Printed Circuit Board
PDMS	Polydimethylsiloxane
PEF	Power Efficiency Metric
PPM	Pulse Position Modulation
PSD	Power Spectral Density
PSK	Phase Shift Keying
PTE	Power Transfer Efficiency
PWM	Pulse Width Modulation
PZT	Piezoelectric
RF	Radio Frequency
SAR	Successive Approximation Register
SD	Standard Deviation
SNR	Signal-to-Noise Ratio
SR	Set-Reset
US	Ultrasound
UWB	Ultra-Wideband
VCO	Voltage Controlled Oscillator
WHO	World Health Organization
WPT	Wireless Power Transfer

Chapter 1
Introduction

1.1 The Impact of Technology in Healthcare

The impact of technological advancement became significant and still has an essential role to play in the diagnosis and treatment of diseases. Before the involvement of the technology in the medical area, the diagnosis was made based on the patient's complaint and the physician's physical examination. The accuracy of the diagnosis was mainly determined by how the patient knows himself or herself and how well he or she describes the complaint. Moreover, the patient could be incompatible with the examination. With the progress of the blood and urine test systems, the accuracy of the diagnosis was dramatically increased. The invention and development of medical imaging systems made a notable enhancement in the diagnosis. Thanks to the increase of these technological tools and successful diagnosis rates, the success in treatment improved a lot. This progress was not only in diagnosis but also in the treatment. Generic treatments were applied for the diseases such that the same medication with the same dosage and for the same duration was recommended for certain symptoms. However, nowadays, personalized medicine and personalized treatment are on the rise as a consequence of the developed and cheaper medical technologies which enables the monitoring and evaluation of many parameters of patients. Furthermore, wearable, transcutaneous, and implantable devices can measure the physical and chemical parameters like blood pressure, pulse rate, glucose, and pH levels, and allow better diagnosis and close tracking of how treatment reacts well with the disease.

One of the areas where the technological developments in biomedical applications can be used most effectively is the neurological monitoring systems. Thanks to these systems, the popular and curious question of "How does the brain work?" might be answered and they can be used effectively in the diagnosis and treatment of many diseases. The brain–machine interface (BMI) allows controlling external software or hardware by translating neural information. Besides, BMI can be used

© Springer Nature Switzerland AG 2020
K. Türe et al., *Wireless Power Transfer and Data Communication for Intracranial Neural Recording Applications*, Analog Circuits and Signal Processing,
https://doi.org/10.1007/978-3-030-40826-8_1

for recovering lost functions like walking. Recently, a non-human primate has regained control of its paralyzed leg, thanks to the brain–spine interface [1]. In addition, as brain signals were analyzed better, this technology can be used to understand the cause and the treatment of neurological disorders such as epilepsy or Parkinson's diseases. The development of the closed-loop systems is still going on to detect the abnormalities in neural signals and prevent them by applying short electrical pulses. This book presents an implantable system and its building blocks for monitoring epilepsy disease as a case study for intracranial neural recording applications.

1.2 Epilepsy and Surgical Treatment

Epilepsy is a chronic discomfort that affects many people throughout the world. Repeated crises can cause reluctant movements involving a part of the body or the entire body. In some cases, there is even a loss of consciousness. According to the World Health Organization (WHO), over 50 million people around the world suffer from epilepsy and this makes it one of the most common diseases [2]. The major form of treatment is long-term drug therapy to recover ~70% of patients [3]. Epilepsy surgery is an alternative for patients that do not respond to drug treatment when a seizure is limited to one area. The probability of becoming seizure-free after surgery is about 65% of those with a lesion identified on MRI, compared with 35% of the non-lesional patients [4]. The main explanation of the failure in surgery is an imprecise localization of foci, which makes the presurgical evaluation to identify the epileptogenic region very crucial.

1.2.1 Neural Activity Monitoring Methods

Continuous monitoring of neural activity is commonly used for the diagnosis of epileptic zones. The measurement methods of the neural signals can be classified into four groups based on the used electrodes, interested signal, and the placement of the sensors.

- Electroencephalogram (EEG) is a non-invasive procedure and the easiest and most commonly used clinical technique in recording electrical activities with flat electrodes placed along the scalp. EEG measures the superpositions of the changes of the field potentials produced by the neural activities. The measurable power spectrum of the EEG signal is limited to 100 Hz and the signal amplitude changes from 5 to 300 μV [5].
- Electrocorticography (ECoG) or intracranial EEG (iEEG) is another kind of EEG and performed by placing the array of electrodes onto the cortex. This procedure is invasive and requires a craniotomy. Compared to EEG, removing the skull and

scalp medium between brain and electrodes provides higher temporal and spatial resolution in ECoG monitoring. ECoG has a broader bandwidth of up to 200 Hz and higher amplitude, which can reach 5 mV [5]. These superior characteristics make ECoG preferable for the presurgical evaluation.

- Local field potentials (LFP) can be measured by using single or multiple needle -shaped electrodes, which can reach the deeper parts of the brain so that it is referred to as a depth recording. This kind of electrodes is also called as penetrating electrodes. Unlike EEG and ECoG, LFP measurement is a highly invasive operation. The frequency content is very similar to ECoG but amplitudes as high as 1 mV [6].

- Spike recordings extract the high-frequency content of individual neuron action potential by using the penetrating electrodes. Since needle-shaped electrodes provide closer distance to the neuron, spike recordings achieve better signal amplitude, which can reach 500 μV. Compared to the other monitoring methods, spike recording has a broader spectrum from 100 Hz to 7 kHz [7].

1.2.2   Macro and Micro-Electrodes for Intracranial Recordings

The presurgical evaluation starts with the non-invasive electroencephalogram (EEG) measurements. Detected abnormal electrical changes help to localize the epilepto-genic zone. However, in some patients, non-invasive measurements are inconclusive because EEG signals are distorted by the tissue such as the skull, veins, meninges, and skin. Besides, the signal activity needs to take place on a large cortical surface ($\sim$6 cm^2) in order to be recognizable by the scalp electrodes [8]. Considering the limits of the EEG electrodes, the intracranial electrodes have advantages because they provide higher spatial and temporal resolutions so that they give a much clearer view of small loci of activity, which is difficult to see on the scalp [9] (Fig. 1.1).

Traditionally, widely spaced (5–10 mm) large electrodes (1–10 mm diameter) for the intracranial measurements have been used [11]. Figure 1.2 depicts the placement of electrodes for iEEG measurements. However, this practice has been adopted because of EGG tradition and the limits of sensor technology rather than the knowledge of the human brain.

Recent studies showed that the temporal and spatial resolution of the measure-ments can be improved further by replacing macro-electrodes with micro-electrodes ($<$100 μm diameter) [13]. The smaller dimension of the electrodes enables the detection of the neural activities localized in sub-millimeter scale regions, which increases spatial resolution. Besides, physicians are able to detect the changes earlier since macro-electrodes require a sufficient population of synchronous activity. This property of micro-electrodes improves the temporal resolution of the measurement. It has been shown that thanks to the better spatial resolution of micro-electrodes, the recorded signal by the adjacent electrodes can be uncorrelated even they are in the vicinity of each other. This feature enables better localization of the epileptogenic zone. Furthermore, detecting the changes in the sub-millimeter region

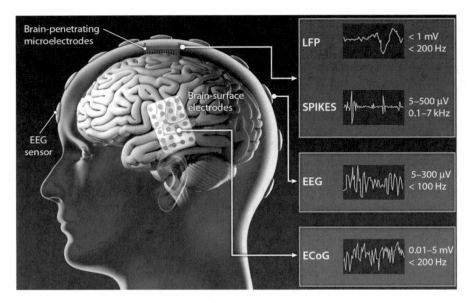

Fig. 1.1 Neural activity monitoring methods. Adapted from [10]

Fig. 1.2 Placement of electrodes for iEEG measurement. Adapted from [12]

before spreading to the neighboring region makes it possible to predict the oncoming seizures. The disadvantage of using micro-electrodes is its limited coverage because of its narrow field of view. This can be handled by increasing the number of electrodes, which will result in higher system complexity with hundreds of channels.

1.2.3   Challenges in Intracranial Recordings

As it is shown in Fig. 1.2, the measured signals by the electrodes are transferred to an external unit through cables. In this unit, the signals are amplified, quantized, and analyzed to present meaningful data to the physicians.

The advantages of intracranial surgery increase the success rate of the presurgical evaluation; however, it is a highly sensitive methodology. The procedure requires a craniotomy and after the placement of electrodes, the skull and scalp are covered but not entirely closed because of the wires coming out from the electrodes. Depending on the situation of the patient, the duration of monitoring can hold several days and even several weeks. The operation itself and the situation during the monitoring can cause some complications.

Cerebrospinal fluid (CSF) is the liquid that goes around, protects, and nourishes the brain. The opening for the wires creates a risk of persistent CSF leakage. There is even a risk of infection of the CSF, which may cause meningitis and encephalitis [14]. A headache and low-grade fever are minor postoperation complications [15]. Since this methodology requires to be connected to an external unit, the duration of monitoring is limited, and it affects the localization success rate. In addition to the medical difficulties, being connected to an external machine restricts the movements of the patient and creates discomfort.

Removing the wires in intracranial EEG enables to cover the brain entirely and eliminates or reduces the possible complications. This book proposes an entirely wireless implantable system for iEEG measurements of presurgical evaluation of epilepsy without sacrificing the benefits of conventional iEEG.

1.3   The Goal: Implantable Monitoring System

Thanks to the advances in electronics, sensor systems, and fabrication methodologies, this book proposes a completely implantable wireless system that eliminates the wires and possible complications due to them. This work aims to alter the quality and quantity of extracted information during the presurgical evaluation of epilepsy and improve the patient's comfort by performing the following enhancements:

1. Commercial macro-electrodes will be replaced with commercial micro-electrodes for smaller form factor and better temporal and spatial resolution.
2. The external unit for amplification and digitalization of the measured signal will be miniaturized to be implanted in the brain.
3. The wires going out of the brain will be removed and transfer of the signal measurements will be established by wireless communication.
4. The external base station will be equipped with receiver and decoder functions.

1.3.1 Wireless Implantable Neural Recording Systems

One of the first integrated wireless systems for neural recording was presented by Akin et al. [16]. The system allows the recording of peripheral neural regeneration through a micromachined silicon sieve electrode. These signals are amplified using on-chip amplifiers, multiplexed, and digitized by an 8-bit analog-to-digital converter (ADC). The system included radio frequency (RF) telemetry link for supplying power to the implant and transmitting the acquired data to an external processor. The overall circuit, including the RF interface circuitry, dissipates 90 mW of power and consumes 4 mm × 6 mm of an area.

As a consequence of emerging efforts, miniaturization in the physical dimension of the implantable system and reduction of its power consumption has been achieved while increasing the number of recording channels and extracted information. Bofanti et al. [17] presented a 64-channel system (16 active channels plus 48 "mute" lines) made by a low-noise analog front-end, a single 8-bit ADC, followed by digital signal compression and transmission units. The dedicated amplifiers amplify the measurements from each channel, while a time-division multiplexer (TDM) and an ADC are used for the digitalization of the data. The compression of the raw data is implemented by detecting the action potential spikes and storing up to 20 points for each waveform. The 400 MHz frequency-shift keying (FSK) transmitter with 1.25 Mbps data rate is used to deliver the information. The battery-powered chip occupies a 3.1 mm × 2.7 mm area and consumes 16.6 mW power.

Another multichannel system presented by Lee et al. [18] includes inductively powered 32-channel system. The wireless links in the industrial-scientific-medical (ISM) band at 915 and 13.56 MHz are used for wireless neural recording and inductive powering, respectively. Unlike the conventional method of using ADC block and thanks to an array of high-speed converters, the amplified neural signals are converted to the pulse width modulated (PWM) form and serialized by a time-division multiplexer. The integrated system occupies 4.9 mm × 3.3 mm and consumes 5.85 mW when 12 out of 32 low-noise amplifiers are active.

The work published by Abdelhalim et al. [19] consists of a 64-channel neural recording interface with 64 programmable 16-tap mixed-signal FIR filters that enable precise separation of various bands in the neural spectrum. The 915 MHz FSK/OOK transmitter offers data rates up to 1.5 Mbps. The fabricated 4 mm × 3 mm chip dissipates 5.03 mW from a 1.2 V supply.

A more recent head-mounted wireless multichannel system reported by Gao et al. [20] offers simultaneous access to 96 channels of broadband neural data. The battery-powered system transmits the digitized recordings with an ultra-wideband (UWB) transmitter at a 40 Mbps data rate. The sensor interface consumes 6.4 W from 1.2 V while occupying 5 mm × 5 mm.

1.3.2   Anticipated Challenges

The design of an implanted monitoring system comes with not only the challenges regarding circuit design but also the limitations imposed by being placed inside the body. The content of this book requires to cope with the challenges and find an optimum selection because of trade-offs. The research effort will focus on the following subjects:

- Size: The micro-electrodes will be on the surface of the cortex, and the central electronics will be placed in an opened hole in the skull. Therefore, the primary physical limitation of the system is the size of the implanted electronics.
- Temperature elevation: The consumed power by the implanted electronics causes a temperature increase in the tissue. The low power consumption of the system has to be realized in order to meet the temperature elevation regulations imposed for implanted medical devices.
- Power source: Proper functioning of the implanted device requires a stable energy source. Removal of the wires and miniaturization of the system force to find an optimum way of supplying energy to the circuits.
- Data rate: Replacing macro-electrodes with micro-electrodes decreases the covered area. Increasing the number of electrodes enlarges the monitored area; however, it increases the generated data, which needs to be transmitted wirelessly.
- Biocompatible packaging: The implanted system needs to be packaged such that it will not cause any damage in the body and it will continue to operate as expected. The material and method of the packaging have to satisfy biocompatibility criteria for several weeks during the monitoring.

1.3.3   Research Objectives

The project presented in this book was performed in a collaboration between Radio Frequency Integrated Circuits (RFIC) Group and Microelectronic Systems Laboratory (LSM) in École Polytechnique Fédérale de Lausanne (EPFL). This book will present the system overview, details, and implementation of the circuits performed by the RFIC group. The main objectives of this study can be classified into three main topics:

- to provide energy to the implanted electronics,
- to transmit the collected data wirelessly,
- to realize a biocompatible package of the implanted electronics.

1.4 Book Outline

In Chap. 2, the system overview and specifications of the proposed architecture are discussed in detail.

In Chap. 3, the power source of the implantable system is explained. Possible solutions for implanted systems are discussed, and their advantages and drawbacks are compared. The chosen structure and implemented circuits are described.

In Chap. 4, the data communication between the implanted system and the external base station is presented. Three different data communication circuits and their performance are exhibited.

In Chap. 5, the biomedical packing of the electronics is discussed, and the *in vivo* experiment results are presented.

In Chap. 6, the conclusion of the study and a perspective on future works are given.

References

1. Capogrosso M, Milekovic T, Borton D et al (2016) A brain–spine interface alleviating gait deficits after spinal cord injury in primates. Nature 539:284–288
2. Epilepsy. In: World Health Organization. http://www.who.int/news-room/fact-sheets/detail/epilepsy. Accessed 22 Oct 2018
3. Kwan P, Brodie MJ (2000) Early identification of refractory epilepsy. N Engl J Med 342:314–319
4. Berkovic SF, McIntosh AM, Kalnins RM et al (1995) Preoperative MRI predicts outcome of temporal lobectomy: an actuarial analysis. Neurology 45:1358–1363
5. Webster JG (2009) Medical instrumentation: application and design, 4th edn. Wiley, Hoboken
6. Andersen RA, Musallam S, Pesaran B (2004) Selecting the signals for a brain–machine interface. Curr Opin Neurobiol 14:720–726
7. Najafi K, Wise KD (1986) An implantable multielectrode array with on-chip signal processing. IEEE J Solid State Circuits 21:1035–1044
8. Spencer D, Nguyen DK, Sivaraju A (2015) Invasive EEG in presurgical evaluation of epilepsy. In: The treatment of epilepsy. Wiley-Blackwell, Hoboken, pp 733–755
9. Nair DR, Burgess R, McIntyre CC, Lüders H (2008) Chronic subdural electrodes in the management of epilepsy. Clin Neurophysiol 119:11–28
10. Thakor NV (2013) Translating the brain-machine interface. Sci Transl Med 5:210ps17
11. Engel J, Pedley TA, Aicardi J (2008) Epilepsy: a comprehensive textbook. Lippincott Williams & Wilkins, Philadelphia
12. Bingaman WE, Bulacio J (2014) Placement of subdural grids in pediatric patients: technique and results. Childs Nerv Syst 30:1897–1904
13. Stead M, Bower M, Brinkmann BH et al (2010) Microseizures and the spatiotemporal scales of human partial epilepsy. Brain 133:2789–2797
14. Van Gompel JJ, Worrell GA, Bell ML et al (2008) Intracranial electroencephalography with subdural grid electrodes: techniques, complications, and outcomes. Neurosurgery 63:498–505; discussion 505–506
15. Fountas KN (2011) Implanted subdural electrodes: safety issues and complication avoidance. Neurosurg Clin N Am 22:519–531, vii

16. Akin T, Najafi K, Bradley RM (1998) A wireless implantable multichannel digital neural recording system for a micromachined sieve electrode. IEEE J Solid State Circuits 33:109–118
17. Bonfanti A, Ceravolo M, Zambra G et al (2010) A multi-channel low-power IC for neural spike recording with data compression and narrowband 400-MHz MC-FSK wireless transmission. In: 2010 Proceedings of ESSCIRC, pp 330–333
18. Lee SB, Lee H, Kiani M, et al (2010) An inductively powered scalable 32-channel wireless neural recording system-on-a-chip for neuroscience applications. IEEE Trans Biomed Circuits Syst 4:360–371
19. Abdelhalim K, Kokarovtseva L, Velazquez JLP, Genov R (2013) 915-MHz FSK/OOK wireless neural recording SoC with 64 mixed-signal FIR filters. IEEE J Solid State Circuits 48:2478–2493
20. Gao H, Walker RM, Nuyujukian P, et al (2012) HermesE: a 96-channel full data rate direct neural interface in 0.13 μm CMOS. IEEE J Solid State Circuits 47:1043–1055

Chapter 2
Implantable Monitoring System for Epilepsy

2.1 System Overview

The proposed system is mainly composed of different electronic blocks dedicated to the measurement of the brain signals, amplification and digitalization of them, and the transmission of the data to an external base station. The whole measurement chain requires an uninterrupted energy source and proper power management circuits to operate during the presurgical analysis. The system-level representation of the proposed system is depicted in Fig. 2.1.

Figure 2.2 illustrates the implantation of the proposed system. The internal base station composed of power management and wireless communication circuits, which are the objectives of this book, is planned to be implanted in the opening on the skull. This hole, namely Burr hole, has a cylindrical shape with a height of skull thickness ($\sim$10 mm) and a diameter of drill size ($\sim$15 mm). The Burr hole determines the size limitations for the proposed implantable system. The peripheral sensor chips which contain micro-electrode arrays (MEAs) and sensor readout circuits such as low-noise amplifiers (LNAs) and analog-to-digital converters (ADCs) are distributed on the cortex. The internal base station and peripheral units are connected with flexible wires in order to provide power to the readout circuits and receive the digitalized neural data, which will be wirelessly transmitted.

2.1.1 Electrodes and Analog Front-End

The acquisition of neural signals improves the understanding of the epilepsy disease and developing new methods for presurgical evaluation. Recent studies showed that the detected high-frequency oscillations (HFO) contain useful information and can be used as biomarkers of epileptogenesis [1]. HFOs are categorized as

© Springer Nature Switzerland AG 2020
K. Türe et al., *Wireless Power Transfer and Data Communication for Intracranial Neural Recording Applications*, Analog Circuits and Signal Processing,
https://doi.org/10.1007/978-3-030-40826-8_2

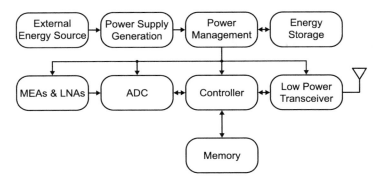

Fig. 2.1 Block diagram of the proposed implant electronics for neural monitoring

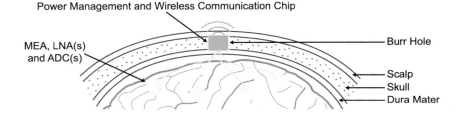

Fig. 2.2 Illustration of the proposed intracranial neural monitoring system

neurological oscillation with a frequency higher than 80 MHz. At present, HFOs are further subclassified into ripples (80–250 Hz) and fast ripples (250–600 Hz) [2]. The frequency spectrum of the neural activity determines the monitoring methodology and the type of electrodes. Considering the interested frequency range, iEEG (or ECoG) shows better performance compared to the conventional EEG methodology. In comparison with macro-electrodes, micro-electrodes provide higher spatial and temporal resolution and provide a better signal-to-noise ratio (SNR) by being placed closer to the cell membrane. Consequently, in this study, it has been decided to use intracranial measurements with arrays of micro-electrodes.

For intracranial measurements, there are mainly two types of electrodes: needle-shaped penetrating electrodes and flat-shaped surface electrodes. Figure 2.3a shows silicon-based Utah penetrating micro-electrodes. Although such electrodes perform well in acute recordings, for chronic monitorings, tissue reaction to the electrode can lead to difficulties such as mechanical trauma, long-term inflammation, and the glial scar formation [5, 6]. Therefore, for long-term measurements, surface electrodes that do not penetrate the neural tissue are preferable. Nowadays, there is a huge effort for the development of flexible, high-density multielectrode arrays (MEAs) as in Fig. 2.3b. Advanced microfabrication processes and materials such as carbon nanotubes, graphene, and poly (3,4-ethylenedioxythiophene) (PEDOT) enable higher resolution, lower impedance, and better biocompatibility [4, 7]. In

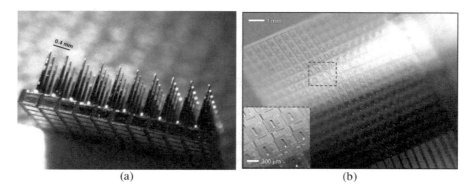

Fig. 2.3 (**a**) Photograph of silicon-based Utah Electrode Array. Adapted from [3]. (**b**) Photograph of a 360 channel high-density active micro-electrode array. Adapted from [4]

this project, a commercially available, flexible polyimide substrate MEA has been selected.

The signal collected from the cortex is very weak, typically between 10 to a few hundred μV [8]. Measurement by the micro-electrodes contains some noise because of the firing of other neurons in the neighborhood. Furthermore, the electrodes introduce noise because of its finite impedance [9]. There is a great effort in the design of electrodes to minimize noise without sacrificing signal integrity.

The collected weak neural signal from each channel in an MEA with a high number of electrodes needs to be amplified before processing and transmitted to the external unit. The low-noise amplifier (LNA) is used in order to amplify the signal with minimum SNR degradation. All the channels by time multiplexing can share a central LNA; however, switches between the electrodes and LNA will cause additional noise in the measurement. Therefore, in the presented project, it was decided to implement an LNA for each channel. A high number of channels result in a high number of LNAs, which cost silicon area and electrical power. Accordingly, the trade-off between noise performance, power consumption, and area requirements has to be intensely studied. In order to compare the noise specifications of LNAs, the well-established noise efficiency factor (NEF) was introduced as a metric [10]. NEF normalizes the input-referred noise of the amplifier to the input-referred noise of a single BJT, which consumes the same current. The NEF includes the total equivalent input noise, total current drain in the system, and the bandwidth of the amplifier. However, the NEF is insufficient to describe the power efficiency of the LNA. To mitigate this issue, the supply voltage of the LNA was added to NEF and called as the power efficiency metric (PEF) [11]. A comparative study of low-noise amplifiers for neural applications showed that the capacitive feedback network (CFN) approach achieves the best performance in terms of area and power consumption for a given input-referred noise specification [12]. In this project, a two-stage fully differential capacitive-coupled low-noise amplifier (CC-LNA) is used for the amplification of neural signals.

2.1.2   Analog to Digital Converter and Compression

The amplified neural signals have to be followed by an analog-to-digital converter (ADC) to digitize signals to a digital representation. The continuous-time
measurement is sampled with a particular frequency, and its amplitude at each
sample is converted to a digital value. According to the Nyquist theorem, a signal
has to be sampled with the double of its frequency for full reconstruction of the
waveform [13]. To record high-frequency oscillations during intracranial EEG, the
amplified signals need to be sampled at least at 2 kHz [2].

The performance of an ADC is generally represented by its SNR, resolution
(or the number of bits), and sampling rate. There are several different ADC
architectures in the literature; however, this project requires energy and area efficient
design. Accordingly, successive-approximation register (SAR) ADCs are used as
they exhibit ultra-low power consumption for moderate resolution and low sample
rates [14].

For the systems with a high number of channels as in the goal of this book, the
selection of the digitalization technique plays an essential role in the performance
of the system. There are two main strategies that can be adopted. The first scenario
is using a dedicated ADC for the digitalization of the signal from each channel [15].
The second option is to utilize a single shared ADC by using an analog multiplexer
which connects the outputs of the LNAs to the input of ADC in a time-interleaved
manner [16]. Although the first method provides better resolution since there is no
SNR degradation caused by the multiplexer, the second approach saves silicon area
with the cost of losing resolution.

From the system-level perspective, the determination of the specifications of
the ADC affects not only the area and power requirement of the ADC but also
the selection, performance metrics, and cost of the wireless transmitter block. For
example, increasing the resolution of the ADC by one bit results in a larger area
and higher power consumption of ADC as well as a higher number of bits to be
transmitted. The payload of the wireless transmitter can be reduced by applying
data compression techniques without losing vital information. The compressions
techniques become popular with the increasing number of channels in implantable
systems [17–19]. Compression exchanges the cost of transmitting a high amount of
data with off-line signal reconstruction in the external base station. The compression
can be applied either in the analog domain by placing the compression block
between analog front-end and ADC [20] or in the digital domain using digital signal
processing (DSP) techniques after digitalization by ADC [21].

2.1.3   Controller Unit

In order to operate efficiently in different situations and cope with the difficulties,
the implanted electronics need to be configurable from the outside. The need for
the number of active channels or sampling rate of the ADCs can be different at

the beginning and the proceeding stages of the presurgical analysis. Moreover, designing implantable electronics requires some precautions for the introduced effects of packaging and the tissue. For instance, there might be a shift in the wireless transmission frequency due to packaging and this shift can be adjusted by proper adjustments of the transmitter parameters. For such kind of changes in the operation or settings of the implanted system, there is a need for a controller unit. Furthermore, in case of using multiple ADCs in the system, there is a need for the serialization of the data from multiple channels before wireless transmission and controller unit is in charge of this operation.

In the proposed system, there is a communication channel from the external base station to the implant, namely downlink communication, to program the controller unit. Depending on the situations and phase of the presurgical analysis, the controller unit enables the following operations:

- activation/deactivation of channels,
- selection of the sampling rate of ADCs,
- fine-tuning of the transmitter frequency,
- selection of the operation mode of the transmitter,
- selection of compression ratio,
- serialization of digital data from multiple ADCs.

2.1.4 Energy Source of the Implanted System

In an implanted system, one of the main issues is the energy source, which enables the operation of the system. The nature of implantation and the available physical dimension of the application imposes limitations on the electrical power source of the system. The energy needed by the implanted electronics can be provided by one or a combination of the following methods:

- medical grade batteries,
- ambient energy harvesting,
- wireless power transfer.

The application determines the appropriate solution. Implantable electronics for medical applications typically require large energy reservoirs to operate reliably over long periods. Harvesting energy from nearby energy sources is an alternative approach to extend implant life and, with sufficient available power, to allow the implant to operate autonomously. For biomedical applications, there are different types of energy harvesters, such as piezoelectric [22], thermal [23], light [24], and infrared light [25] with a power density of microwatts per cm^2. Although the energy harvested by these types of devices is limited, they can be used for ultra-low power implants. Mercier et al. demonstrated an electronic system able to extract a minimum of $1.12\,nW$ from the endocochlear potential (EP) of a guinea pig for up to $5\,h$, enabling a $2.4\,GHz$ radio to transmit the measurement of the EP every $40–360\,s$ [26].

Batteries should either be replaced by additional surgery or recharged for long-term implants. Charge capacities of the batteries and required power for a particular time of operation should also be considered while selecting the battery, as well as the size restrictions of the implant. As an example, Miranda et al. present a wireless biomedical system for recording and transmitting the neural activity of the brain with 32 channels. The power consumption is low enough to operate continuously for 33 h, using two 3.6 V/1200 mAh Li-SOCl$_2$ batteries [27]. A rechargeable battery can be used in order to extend the duration of the operation. However, recharging an implanted battery requires wireless power transfer.

Wireless power transfer solutions are appropriate for applications such as neural monitoring, which requires a long-term operation and consumes in the order of milliwatts of power. Remote powering solutions can be analyzed in two sections concerning the distance between the external and the implanted units: near field and far-field. The boundary between the near field and far-field is defined by d= $\lambda/2\pi$, where d and λ are the distance and the wavelength of the signal, respectively. Long-distance (a few meters) remote powering solutions generally exploit far-field properties, more explicitly radiation properties, of antennas at several hundreds of MHz frequencies [28]. Therefore, they are more suitable for applications that require high mobility. On the other hand, short-distance (a few centimeters) remote powering solutions employ reactive coupling techniques such as capacitive and inductive at several MHz frequencies. As the names imply, capacitive coupling is a result of electrical coupling, whereas inductive coupling is of magnetic coupling. Capacitive coupling requires a dielectric medium that allows strong coupling and is more sensitive to distance variations. On the other hand, the inductive coupling method exploits the mutual inductance between coupled inductors. In the literature, there are numerous examples of wireless power transfer by means of electromagnetic (EM) radiation [29, 30], magnetic coupling [31–33], ultrasonic coupling [34], and infrared radiation [35]. For a chosen wavelength, if the distance between the coils is smaller than d, the magnetic coupling gives more efficient wireless power transfer [36]. Accordingly, magnetic coupling is preferable for powering the implanted devices. It is fair to claim that EM radiation and magnetic coupling based systems dominate the literature, especially for neural implant powering applications. Recently, Lee et al. have presented an inductively-powered wireless integrated neural recording system for wireless and battery-less neural recording from freely behaving animal subjects inside a wirelessly powered standard homepage [37]. The proposed system consumes 51.4 mW and an inductive link at 13.56 MHz powers it.

For several biomedical applications such as hearing aids and pacemakers, batteries occupy a significant amount of volume. However, the volume allocated for a neural implant is minimal compared to these applications. Moreover, the neural recording applications consume a higher amount of power and this property reduces the duration of the operation. Considering the continuous power demand of the neural implants aiming for continuous data transmission and the estimated power budget, current ambient energy harvesters are found to be insufficient to fulfill this task. The combination of a thin Li-ion battery and wireless power transfer link utilizing inductive coupling as a power source of the implanted system is a good solution since the distance between the implant and external unit is in the order of millimeters (human scalp thickness ~10 mm) and sending the required power to the implant is feasible with current technology.

2.1.5 Wireless Data Transmission

In addition to the neural data acquisition and processing, a communication channel from the implant to an external base station, namely uplink communication, is required to transmit digitized neural data to the external world. Real-time data transfer entails a wireless data transmission at the same rate as the acquired brain signals. If a memory block is used in order to reduce the need for continuous data transmission, it requires communication with a higher data rate than the acquisition rate. As a result, the communication data rate is, in fact, determined mainly by the quality and quantity of the recording channels. The quality of the channels is measured via resolution, sampling frequency, and bandwidth. If we consider epilepsy monitoring applications specifically, the required data rate is in the order of a few megabits per second (Mbps).

The proposed epilepsy monitoring system in this project requires both uplink and downlink communications. The downlink communication is needed for data transfer from the external base station to the implant in order to configure sensor parameters and processing parameters such as the sampling rate of resolution of ADC. Since the downlink communication is only used for setting the system parameters, there is no need for a high data rate communication. It is sufficient to design a downlink receiver at the implant side that can communicate with the data rate of 10 kbps for this kind of application. However, for uplink communication, very high data rate communication is demanded since the number of monitoring channels and their sampling rate is increasing. The multiplication of the following quantities calculates the total data rate:

- number of micro-electrode arrays,
- number of electrodes in each array,
- ADC sampling rate,
- ADC number of bits,
- compression factor.

For the neural monitoring application with tens of electrodes, uplink communication should provide at least 10 Mbps data rate. Accordingly, the design of an uplink transmitter is a challenging factor in the application. The payload of the upload transmitter can be reduced by applying data compression techniques, which costs processing power and silicon area. This trade-off has to be analyzed deeply for the optimum operating point.

Wireless data communication solutions can be classified into two groups: data communication on the powering line by changing parameters of the wireless power transfer link or employing a dedicated transceiver on both parts. Downlink communication can be directly performed by modulating the signal source in amplitude, frequency, or phase. Uplink communication performed by perturbing the characteristics of the power line is called load modulation for magnetic coupling based power transfer links [38] and backscattering for electromagnetic radiation based links [39]. Using a dedicated transceiver isolates power and data transmission channels, allowing these two links to be designed independently [40]. A compromise between these two solutions can be formulated concerning the power budget and data rate requirement of the recording application. Moreover, additional components such as antennas occupy a non-negligible volume that may violate size restrictions. In both cases, the selection of operating frequencies has to be carried out in careful consideration of the bandwidth requirement imposed by the data rate of the application.

Downlink communication is decided to be built on the inductive power transfer link. The powering signal is modulated by the configuration data of the controller unit and the change in the powering signal is demodulated in the implant. For this application, amplitude shift keying (ASK) type modulation is decided. The powering signal amplitude is changed without disturbing wireless power transfer and data is demodulated by implanted envelope detection circuits. Since the inductive link requires a close distance between the implant and the external base station, downlink communication and programming controller unit is only possible when the external unit is in the proximity of the implanted system.

For uplink communication, three different methods are used in order to be adaptable for different situations depending on the data rate and transmission distance.

- *Communication on powering line*: A load modulation technique is used for the transmission with a few Mbps data rate. A resonant circuit has to be created for efficient power transmission with inductive coupling. A parallel or a series connection of a capacitor with the implanted inductive coil constitutes the resonant circuit. The shift in the resonance frequency in the implanted system causes a change in the operating conditions of the external base station. In the proposed method, the uplink data modulates the resonance frequency by changing the parallel capacitance value and the data is demodulated in the external base station by following the changes in the voltage over the power transmitting coil. Using powering coils for the uplink transmission eliminated the need for an additional antenna; however, the data rate is limited to a few Mbps.

- *Communication with dedicated narrowband transmitter*: For higher transmission rates, for which load modulation is not sufficient, there is a need for a dedicated transmitter. An on-off keying (OOK) transmitter working at the MedRadio band is adopted for the data rates of about 10 Mbps. MedRadio spectrum is used for diagnostic and therapeutic purposes in implanted medical devices as well as devices worn on a body [41]. The transmitter is composed of an oscillator and an off-chip antenna. The digitized neural recording modulates the operation of the oscillator by turning it on or off. The antenna radiates waves when the data bit is "1" and stops radiation when it is "0." The receiver unit in the external base station demodulates the radiation pattern. Compared to the communication on the power line, this method requires an additional antenna; however, it is capable of sending information at higher speed and for longer distances.

- *Communication with dedicated ultra-wideband transmitter*: For very high transmission rates around 100 Mbps or more, the narrowband transmitters become inadequate because of their capabilities and the communication regulations. Federal Communication Commission (FCC) authorized the unlicensed use of ultra-wideband (UWB) in the frequency range from 3.1 to 10.6 GHz with maximum power spectral density (PSD) of −41.3 dBm/MHz [42]. For high data rate communication, an impulse radio ultra-wideband (IR-UWB) is proposed in the authorized band. The designed IR-UWB works with OOK modulation and creates pulses with very short duration for only bit "1"s while does nothing for bit "0"s. Like a narrowband transmitter, an IR-UWB requires an extra antenna; even so, the transmission distance is minimal due to the regulations and the high loss in the tissue at high frequencies [43]. Therefore, this method is only available when the external station is close to the implant.

2.2 System Specifications

The targeted wireless neural recording system specifications are determined with a strong collaboration with the Department of Neurology at Bern University Hospital and summarized in Table 2.1. Figure 2.4 demonstrates the detailed block diagram of the proposed implant electronics for neural monitoring. As aforementioned, the high-frequency oscillations (80–600 Hz) contain useful information for the localization of the epileptogenic zone. The bandwidth of the neural recording amplifier and ADC are determined accordingly. For better presurgical analysis and localization of the responsible region for epileptic seizures, a 10-bit resolution is targeted for 64 recording channels. Setting the sampling rate to 4 kilo-samples per

Table 2.1 System specifications for implantable neural monitoring system

Design parameter	Value
Number of channels	64
Bandwidth of interested signal (Hz)	80–600
ADC sampling rate (ksps)	4
ADC resolution (bits)	10
Uplink data rate (Mbps)	10
Downlink data rate (kbps)	10
Power budget (mW)	10
Powering distance (mm)	10
Implant diameter (mm)	15
Implant height (mm)	10

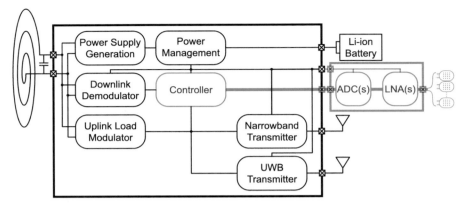

Fig. 2.4 Detailed block diagram of the proposed implant electronics for neural monitoring

second (ksps) yields 2.56 Mbps raw data rate. Considering the future increase in sampling rate and the number of channels, the minimum data rate limit for the uplink transmitter is set to 10 Mbps. A downlink data rate of 10 kbps is decided to be able to send necessary commands to the controller unit in the system.

The intended implantable neural monitoring application also determines the physical limits, which will affect the size and available power budget for the system. The scalp thickness imposes the distance between the implant and external unit, which is around 10 mm. The implanted system is expected to be placed in a cylindrical shaped Burr hole in the skull. The estimated diameter of the hole is 15 mm. The height of the hole is the same as the thickness of the skull, and it is expected to be around 10 mm. Burr hole defines the boundary condition for the physical dimensions of the system. Besides, the ratio of dissipated power to the surface area of the implanted system, namely power outflux density, determines the temperature rise, which may disturb the natural behavior of the cells nearby the implant or may even cause cell death. The total power budget of the system is aimed to be 10 mW by taking into account 0.1 mW consumption for each channel and possible extension in the number of channels. The numerical analysis showed that maximum allowable total power dissipation in a cortical implant of size $2 \times 2 \times 0.5\,\text{mm}^3$ is found to be 4.8 mW [44]. The surface area of the proposed system is more than ten times bigger than the implant used in numerical analysis, which proves that the temperature regulations will be met (Fig. 2.4).

2.3   Summary

In this chapter, the system-level approach to the fully implanted intracranial monitoring of epilepsy has been covered. Solutions and specifications for each building block have been presented. Furthermore, a detailed block diagram of the implanted system has been depicted. The next chapter gives more information about inductive link and power management circuits for remote powering.

References

1. Wu JY, Sankar R, Lerner JT et al (2010) Removing interictal fast ripples on electrocorticography linked with seizure freedom in children. Neurology 75:1686
2. Zijlmans M, Jiruska P, Zelmann R et al (2012) High-frequency oscillations as a new biomarker in epilepsy. Ann Neurol 71:169–178
3. Kim S, Bhandari R, Klein M et al (2009) Integrated wireless neural interface based on the Utah electrode array. Biomed Microdevices 11:453–466
4. Viventi J, Kim D-H, Vigeland L et al (2011) Flexible, foldable, actively multiplexed, high-density electrode array for mapping brain activity in vivo. Nat Neurosci 14:1599–1605
5. Yeager JD, Phillips DJ, Rector DM, Bahr DF (2008) Characterization of flexible ECoG electrode arrays for chronic recording in awake rats. J Neurosci Methods 173:279–285
6. Polikov VS, Tresco PA, Reichert WM (2005) Response of brain tissue to chronically implanted neural electrodes. J Neurosci Methods 148:1–18
7. Mehrali M, Bagherifard S, Akbari M et al (2018) Blending electronics with the human body: a pathway toward a cybernetic future. Adv Sci (Weinh) 5(10)
8. Wattanapanitch W, Fee M, Sarpeshkar R (2007) An energy-efficient micropower neural recording amplifier. IEEE Trans Biomed Circuits Syst 1:136–147
9. Yang Z, Zhao Q, Keefer E, Liu W (2009) Noise characterization, modeling, and reduction for in vivo neural recording. In: Bengio Y, Schuurmans D, Lafferty JD et al (eds) Advances in neural information processing systems, vol 22. Curran Associates Inc., New York, pp 2160–2168
10. Steyaert MSJ, Sansen WMC (1987) A micropower low-noise monolithic instrumentation amplifier for medical purposes. IEEE J Solid-State Circuits 22:1163–1168
11. Muller R, Gambini S, Rabaey J (2012) A $0.013\,mm^2$, $5\,\mu W$, DC-coupled neural signal acquisition IC with 0.5 V supply. IEEE J Solid-State Circuits 47(1):232–243
12. Ruiz-Amaya J, Rodríguez-Pérez A, Delgado-Restituto M (2010) A comparative study of low-noise amplifiers for neural applications. In: 2010 International conference on microelectronics, pp 327–330
13. Nyquist H (1928) Certain topics in telegraph transmission theory. Trans Am Inst Electr Eng 47:617–644
14. Gao H, Walker RM, Nuyujukian P et al (2012) HermesE: A 96-channel full data rate direct neural interface in 0.13 µm CMOS. IEEE J Solid-State Circuits 47:1043–1055
15. Mollazadeh M, Murari K, Cauwenberghs G, Thakor N (2009) Micropower CMOS integrated low-noise amplification, filtering, and digitization of multimodal neuropotentials. IEEE Trans Biomed Circuits Syst 3:1–10
16. Harrison RR, Watkins PT, Kier RJ et al (2007) A low-power integrated circuit for a wireless 100-electrode neural recording system. IEEE J Solid-State Circuits 42:123–133
17. Chen F, Chandrakasan AP, Stojanovic VM (2012) Design and analysis of a hardware-efficient compressed sensing architecture for data compression in wireless sensors. IEEE J Solid-State Circuits 47:744–756
18. Shoaran M, Kamal MH, Pollo C et al (2014) Compact low-power cortical recording architecture for compressive multichannel data acquisition. IEEE Trans Biomed Circuits Syst 8:857–870
19. Wang A, Lin F, Jin Z, Xu W (2016) Ultra-Low Power Dynamic Knob in Adaptive Compressed Sensing Towards Biosignal Dynamics. IEEE Transactions on Biomedical Circuits and Systems 10:579–592

20. Ranjandish R, Schmid A (2018) A sub-µ/channel, 16-channel seizure detection and signal acquisition SoC based on multichannel compressive sensing. IEEE Trans Circuits Syst II: Express Briefs 65:1400–1404
21. Aprile C, Ture K, Baldassarre L et al (2018) Adaptive learning-based compressive sampling for low-power wireless implants. IEEE Trans Circuits Syst I: Reg Pap 65:3929–3941
22. Kwon D, Rincon-Mora GA (2010) A 2-µm BiCMOS rectifier-free AC–DC piezoelectric energy harvester-charger IC. IEEE Trans Biomed Circuits Syst 4:400–409
23. Zhang Y, Zhang F, Shakhsheer Y et al (2013) A batteryless 19µW MICS/ISM-Band energy harvesting body sensor node SoC for ExG applications. IEEE J Solid-State Circuits 48:199–213
24. Ayazian S, Hassibi A (2011) Delivering optical power to subcutaneous implanted devices. In: 2011 Annual international conference of the IEEE engineering in medicine and biology society, pp 2874–2877
25. Goto K, Nakagawa T, Nakamura O, Kawata S (2001) An implantable power supply with an optically rechargeable lithium battery. IEEE Trans Biomed Eng 48:830–833
26. Mercier PP, Lysaght AC, Bandyopadhyay S et al (2012) Energy extraction from the biologic battery in the inner ear. Nat Biotechnol 30:1240–1243
27. Miranda H, Gilja V, Chestek CA et al (2010) HermesD: a high-rate long-range wireless transmission system for simultaneous multichannel neural recording applications. IEEE Trans Biomed Circuits Syst 4:181–191
28. Nikitin PV, Rao KVS, Lazar S (2007) An overview of near field UHF RFID. In: 2007 IEEE international conference on RFID, pp 167–174
29. Chow EY, Yang C, Ouyang Y et al (2011) Wireless powering and the study of RF propagation through ocular tissue for development of implantable sensors. IEEE Trans Antennas Propag 59:2379–2387
30. Ho JS, Kim S, Poon ASY (2013) Midfield wireless powering for implantable systems. Proc IEEE 101:1369–1378
31. Yilmaz G, Atasoy O, Dehollain C (2013) Wireless energy and data transfer for in-vivo epileptic focus localization. IEEE Sensors J 13:4172–4179
32. Sauer C, Stanacevic M, Cauwenberghs G, Thakor N (2005) Power harvesting and telemetry in CMOS for implanted devices. IEEE Trans Circuits Syst I: Reg Pap 52:2605–2613
33. Catrysse M, Hermans B, Puers R (2004) An inductive power system with integrated bi-directional data-transmission. Sens Actuators A: Phys 115:221–229
34. Mazzilli F, Thoppay PE, Praplan V, Dehollain C (2012) Ultrasound energy harvesting system for deep implanted-medical-devices (IMDs). In: 2012 IEEE international symposium on circuits and systems, pp 2865–2868
35. Mathieson K, Loudin J, Goetz G et al (2012) Photovoltaic retinal prosthesis with high pixel density. Nat Photonics 6:391–397
36. Kilinc EG, Dehollain C, Maloberti F (2015) Remote powering and data communication for implanted biomedical systems, 1st edn (2016 edition). Springer, New York
37. Lee S, Lee B, Kiani M et al (2016) An inductively-powered wireless neural recording system with a charge sampling analog front-end. IEEE Sensors J 16:475–484
38. Mandal S, Sarpeshkar R (2008) Power-efficient impedance-modulation wireless data links for biomedical implants. IEEE Trans Biomed Circuits Syst 2:301–315
39. Rao KVS, Nikitin PV, Rao KVS, Nikitin PV (2006) Theory and measurement of backscattering from RFID tags. IEEE Antennas and Propag Mag 48:212–218
40. Pandey J, Otis BP (2011) A Sub-100µW MICS/ISM band transmitter based on injection-locking and frequency multiplication. IEEE J Solid-State Circuits 46:1049–1058
41. Medical Device Radiocommunications Service (MedRadio) (2011). In: Federal Communications Commission. https://www.fcc.gov/wireless/bureau-divisions/broadband-division/medical-device-radiocommunications-service-medradio. Accessed 12 Nov 2018
42. First report and order regarding ultra-wideband transmission systems (2002). Federal Communications Commission

43. Vaillancourt P, Djemouai A, Harvey J, Sawan M (1997) EM radiation behavior upon biological tissues in a radio-frequency power transfer link for a cortical visual implant. In: Proceedings of the 19th annual international conference of the IEEE engineering in medicine and biology society 'magnificent milestones and emerging opportunities in medical engineering' (Cat No97CH36136)
44. Silay KM, Dehollain C, Declercq M (2008) Numerical analysis of temperature elevation in the head due to power dissipation in a cortical implant. In: 2008 30th annual international conference of the IEEE engineering in medicine and biology society, pp 951–956

Chapter 3
Powering of the Implanted Monitoring System

3.1 Introduction

The proposed system in Chap. 2 can be configured by the external base station to operate in different modes, which cause various required power budget during the presurgical analysis. The temperature elevation because of the dissipated power may disturb the natural behavior of the cells nearby the implant or may even cause cell death. Regulations allow a maximum $1\,^\circ$C temperature elevation for body implants [1]. This temperature rise corresponds to $40\,$mW/cm^2 power outflux density [2]. In order to satisfy the regulations and safe operation of the implanted system, the maximum power consumption of the system is set to $10\,$mW.

The specifications and limitations define the minimum area that integrated circuits can occupy. The ratio of the maximum power consumption and the maximum power which can be dissipated determines the minimum area of the implanted system which can be occupied. Neglecting the small surface of the sidewall of the chips and noting power dissipates mostly by the top and bottom surfaces of the chips, the required total minimum surface area corresponds to $12.5\,$mm^2.

3.2 Power Sources for Implanted Systems

Powering is one of the significant challenges for a system designed for continuous monitoring. Especially for a system during the presurgical analysis of epilepsy disease, an uninterrupted power source is essential because of unknown seizure times. Considering not only the high number of micro-electrodes for better resolution and large coverage area but also the amplification, digitalization, and wireless transmission of the neural signal results in an important amount of energy need.

© Springer Nature Switzerland AG 2020
K. Türe et al., *Wireless Power Transfer and Data Communication for Intracranial Neural Recording Applications*, Analog Circuits and Signal Processing,
https://doi.org/10.1007/978-3-030-40826-8_3

Besides, the duration of the presurgical analysis can take several days and even several weeks so that the capacity of the powering solution becomes critical.

In literature, there are several different methods to power up an implanted system. Those approaches can be grouped into three categories:

- implanted batteries,
- ambient energy harvesters,
- wireless power transfer.

All the presented methods have their advantages and drawbacks. Therefore, it is not possible to offer a generic way which shows the best performance in every situation. Moreover, the restrictions and specifications alter with the variety of utilization of the implanted system. Accordingly, the application determines itself as the most appropriate solution.

3.2.1 Implanted Batteries

Batteries have been used in medical applications since 1972 [3]. Pacemakers and cochlear implants are the most widely used medical devices with batteries. In the last years, pacemaker therapy has considerably expanded, exceeding 700,000 implantations annually worldwide, according to the survey in 2009 [4]. In addition to the increasing demand in medical applications, with the advancements in renewable energy, widely usage of mobile devices and electric vehicles, which are expected to be the core of the transportation, the batteries have been receiving a close review. However, the energy storage capacities per mass and per volume are still limited. Tables 3.1 and 3.2 show three kinds of primary (single-use) and secondary (rechargeable) batteries, respectively, with their capacities.

The primary batteries have higher energy capacity than the secondary ones. However, the capacities are still limited when the power budget and physical limits of the project are considered. The implanted system is expected to be placed in a cylindrical shaped Burr hole in the skull. The estimated diameter of the hole is 15 mm. The height of the hole is the same as the thickness of the skull, and it is expected to be around 10 mm. When it is assumed that the battery occupies the half

Table 3.1 Energy density and voltage of three primary battery chemistries [5]

	Zinc-air	Lithium	Alkaline
Energy density (J/cm^3)	3780	2880	1200
Voltage (V)	1.4	3.0–4.0	1.5

Table 3.2 Energy density and voltage of three secondary battery chemistries [5]

	Li-ion	NiMHd	NiCd
Energy density (J/cm^3)	1080	860	650
Voltage (V)	3.0	1.5	1.5

Table 3.3 Available energy
and estimated duration of
operations for batteries placed
in Burr hole

	Zinc-air	Li-ion
Energy (J)	3402	972
Estimated operation time (h)	94.5	27

of Burr hole, there is $0.9\,cm^3$ available volume and corresponding available energy
for cells with Zinc-air and Li-ion compositions are tabulated in Table 3.3.

The maximum operation duration for a primary battery is 94.5 h and after that
time, it requires another surgery to replace the battery. Considering the needed
time for presurgical evaluation, multiple operations are expected to complete the
analysis and those surgeries diminish the advantages of iEEG over EEG. Although
the secondary batteries cannot be used as the only energy source of the system
because of their low capacities, they can be a part of a hybrid solution with energy
harvesting or wireless power transmission systems.

3.2.2 Ambient Energy Harvesters

The power harvesting methods convert the different forms of available energy to
electrical power. On the contrary of the batteries, the primary performance metric
for the energy harvesters is power density because the energy conversion efficiency
depends on the instantaneously available source. Moreover, the nature of the source
may force to use secondary storage. Table 3.4 summarizes different kinds of ambient
energy harvesters and their power densities.

The usage of the energy harvesters in implanted biomedical systems is challenging mainly because of two reasons; firstly, the limited availability of the source,
and secondly, the limited space for implants. Nonetheless, there are examples of
piezoelectric [6], thermal [7], light [8], and infrared light [9] harvesters for ultra-low
power implants. Although the energy harvesting is not sufficient for the proposed
application, decreasing power demands of implants and better approaches for energy
harvesters may result in more extensive usage in monitoring and treatment systems.

3.2.3 Wireless Power Transfer

The powering of implantable systems can also be performed by wireless power
transmission (WPT) methods. In the literature, there are numerous examples of
wireless power transfer by means of electromagnetic (EM) radiation [10, 11],
magnetic coupling [12–14], ultrasonic coupling [15], and infrared radiation [16].
All these methods require one power transmitter and a receiver, respectively, outside
and inside the body. The power transmitter converts the electrical power to a carrier
wave, and the receiver unit converts the carrier wave back to electrical energy.

Table 3.4 Summary of various potential power sources for ambient energy harvesters [5]

	Power density (μW/cm^3)	Secondary storage needed
Solar (outside)	15,000[a]	Usually
Solar (inside)	10[a]	Usually
Temperature	40[a,b]	Usually
Human power	330	Yes
Air flow	380[c]	Yes
Pressure variations	17[d]	Yes
Vibrations	300	Yes

[a]Denotes sources whose fundamental metric is power per square centimeter rather than power per cubic centimeter
[b]Demonstrated from a 5 °C temperature differential
[c]Assumes air velocity of 5 m/s and 5%
[d]Based on a 1 cm^3 closed volume of helium undergoing a 10 °C temperature change once per day

Compared to batteries and ambient energy harvesters, wireless power transfer is not an energy source itself, but it is a technique for delivering energy. Accordingly, the characterization of the remote powering regarding its energy or power density is not possible. The challenging aspects of wireless power transmission are provided power, the efficiency of power delivery, and the transmission distance. The application determines the transmission distance and limits the achievable efficiency. However, increasing the power level in the transmitter side increases the delivered power. This property of WPT presents a higher available power budget than batteries and energy harvesters even though there are some regulations on the power transferring mediums. For instance, Lee et al. have presented an inductively powered wireless integrated neural recording system for wireless and battery-less neural recording from freely behaving animal subjects inside a wirelessly powered standard homecage [17]. The proposed system consumes 51.4 mW and an inductive link at 13.56 MHz powers it.

The choice of the method depends on the trade-offs between many parameters such as

- required power by the implant,
- the distance between power source and implant,
- and power transmission environment.

Among powering solutions of the implanted electronics, wireless power transfer stands out as the most appropriate one for high performance and long-term monitoring systems. However, wireless power transfer requires an external receiver all the time. In order to increase the comfort of the patient, a hybrid solution composed of wireless powering and a secondary battery is decided to be used in the system. Thanks to the battery, the patient will be able to continue his or her daily life without being externally connected to any device, and the battery will be recharged by the help of wireless powering.

3.3   Wireless Power Transfer (WPT)

Wireless power transfer or remote powering in biomedical applications can be grouped based on the schemes used in transmission, as shown in Fig. 3.1. Near-field (NF) coupling, far-field (FF) radiation or radio-frequency (RF) transmission, and ultrasound (US) are the various methods to power the implanted system. The boundary between the near-field and far-field is defined by $d = \lambda/2\pi$, where d and λ are the distance and the wavelength of the signal, respectively. Near-field coupling is subdivided into two by the used techniques, namely magnetic field and electric field coupling.

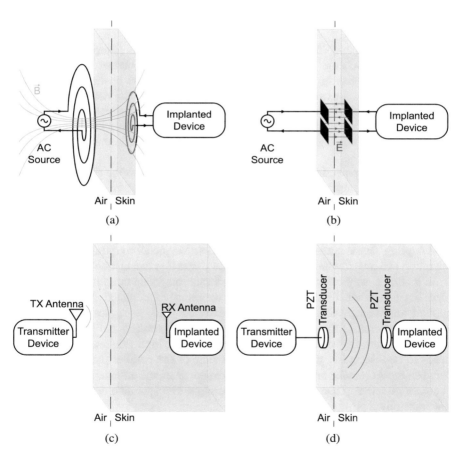

Fig. 3.1 Wireless power transfer strategies for implantable bioelectronics. (**a**) Schematic of inductive coupling method. (**b**) Schematic of capacitive coupling method. (**c**) Schematic of far-field method. (**d**) Schematic of ultrasound method

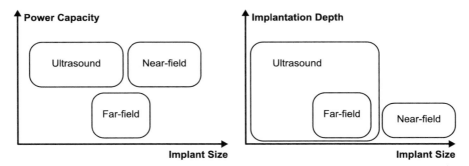

Fig. 3.2 Comparative illustrations of applicable regimes for various WPT schemes in biomedical applications

Each approach has its own merits and drawbacks. The criteria for choosing the appropriate scheme are mainly WPT range, power budget needed by the implant, and the size of the implant. Figure 3.2 shows the abstract view of the applicable regimes for various remote powering methods. While near-field coupling is superior in the transmission of high power for short-range, ultrasound becomes a better choice for deeper implantations. Additionally, the far-field scheme is limited regarding power and range because of attenuation in the tissue at the high-frequency band.

The near-field powering using a magnetic field is based on the placed in parallel two coupled inductors, as shown in Fig. 3.1a. An applied alternating current to the transmitter coil generates a time-varying magnetic field which induces an alternating current in the receiving coil. The power transfer efficiency (PTE), the ratio of received power to the consumed power by the transmitter, depends strongly on the coupling factor between the two inductors. Magnetic coupling benefits from operating at a low-frequency range where the tissue loss can be neglected.

In a similar way of power transmission in the magnetic coupling, two parallel plate capacitors are used for the electric coupling, as depicted in Fig. 3.1b. An applied alternating voltage between the external pairs generates very low displacement currents. The tissue limits the distance between the transmitter and receiver patches, as well as the delivered power. The high frequency of operating needs to be selected to decrease the impedance of the parallel plate capacitances between the AC source and implanted device; however, the attenuation of the tissue becomes effective. Therefore, remote powering with capacitive coupling provides low transmitted power.

The far-field power transmission uses electromagnetic (EM) waves for carrying the power. The transmitter antenna produces the EM waves and current is generated across the terminals of receiver antenna by the radiated waves. The far-field transmission is applicable for the deeply placed implants using smaller antennas compared to those used in near-field delivery. However, the far-field EM channel suffers from tissue losses, likewise capacitive coupling since the link operates at high frequencies. For this reason, the deliverable power to the implant is limited.

The ultrasound power transfer method utilizes the sound waves as the carrier of energy. Similar to the EM transmission, an external piezoelectric (PZT) transducer converts electrical energy to ultrasound waves. Another PZT transducer receives the waves and transforms it into electrical power, as illustrated in Fig. 3.1d. Since ultrasound transmission uses low frequency, it is exposed to low attenuation of tissue, likewise inductive coupling. Moreover, the Food and Drug Administration (FDA) permits an intensity of $7.2 \, \mathrm{mW/mm^2}$ for ultrasound applications [18], which is about two orders of magnitude higher than the EM radiation limit [1]. Taken together, these properties indicate that the ultrasound power delivery is suitable for high power requested applications that are deeply implanted.

The proposed system in this book is planned to be placed in a Burr hole. Therefore the distance between the external station located over the head and the implanted system is considered as a near-field. Taking into account the expected power consumption of the system and design simplicity of the inductive link, the magnetically coupled wireless energy transmission becomes a favorable solution.

3.4   WPT with Magnetic Coupling

3.4.1   Inductive Link

The mechanism of magnetic coupling can be briefly explained using Faraday's law of induction. This law states that any change in the magnetic environment of a coil of wire will induce a voltage. This change in WPT systems based on magnetic coupling is produced by a time-varying current in another coil that is present in the vicinity. Figure 3.3 shows the two coils for inducing a voltage magnetically. It requires other building blocks for converting induced voltage to a stable DC power supply.

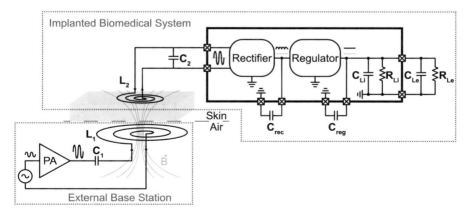

Fig. 3.3 The building blocks of a magnetically coupled WPT system

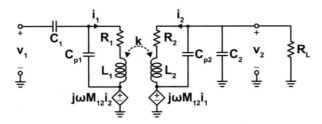

Fig. 3.4 A simplified model of 2-coil resonant inductive link

From a DC power source, the power amplifier (PA) generates sinusoidal current pulses and injects them into the base-station inductor L_1. Due to time-varying current in L_1, changing magnetic flux inside the coil L_2 induces an AC voltage. The capacitance C_1 resonates with L_1 and the capacitance C_2 resonates with L_2 at the same operation frequency to maximize the induced voltage. The rectifier provides a DC voltage from induced voltage and finally a constant and stable voltage is generated by a voltage regulator to supply the signal processing circuits of the remotely powered side. The equivalent on-chip and off-chip loads of remote powering circuits are represented by R_{Li}, R_{Le}, C_{Li}, and C_{Le}.

Figure 3.4 illustrates the simplified model of a 2-coil resonant inductive link. L_1 and L_2 define the inductances created by two coils and the losses of the coils are modeled by R_1 and R_2. C_{p1} and C_{p2} indicate the corresponding parasitic capacitances between the turns of coils. The inductors are brought to the same resonance at ω_0 with appropriate C_1 and C_2 capacitances. Resistor R_L corresponds to the load, which models the power consumption by the rest of the implanted system. The coupling between the 2-coil can be quantified by the coupling coefficient, k, which is modeled by the mutual inductance between the inductors, M_{12}. The relationship between the coupling coefficient and mutual inductance is expressed by

$$k = \frac{M_{12}}{\sqrt{L_1 L_2}}. \tag{3.1}$$

The power transfer efficiency (PTE) of the link in Fig. 3.4 is derived in [19]:

$$\eta_k = \left(\frac{k^2 Q_1 Q_2'}{1 + k^2 Q_1 Q_2'} \right) \left(\frac{Q_2}{Q_2 + Q_L} \right) \tag{3.2}$$

where $Q_k = \omega_0 L_k / R_k$ for $k = 1, 2$, $Q_L = \omega_0 R_L C_2$, $Q_2' = Q_2 Q_L / (Q_2 + Q_L)$ and ω_0 is the operation frequency. Equation 3.2 shows that efficiency is constituted of two mechanisms. The first part represents the losses associated with two inductors and the second part depicts the power division between the parasitic resistance of the second coil and the load. The electrical model and equations show that the efficiency depends on the coupling coefficient (k) between coils and the quality factors (Q) of the coils. This requires a geometric design optimization of spiral coils on a printed circuit board (PCB). The implant size and separation between the implant and base station is generally limited by the targeted application. However, there is still room for optimization of spiral shape, width, spacing, and number of turns. Moreover, the parameters of the external coil are more flexible and implementation of the WPT system can be various for different specific applications.

The dimensions of the implanted coil and the efficiency of WPT are the major limitations in designing the coils for remote powering. As it is shown in Eq. 3.2, there is a direct relationship between the delivered power to the load and the efficiency. However, in the proposed system, the equivalent load of the inductive link is variable depending on different conditions such as the sampling rate of ADC, the number of active channels, and the type of data transmission. The variation in the load power requires an additional approach for keeping PTE maximum for different activity rates.

A modified version of an inductive link with 4-coil instead of 2-coil has been introduced for 2 m remote powering [20], and the structure was adapted for implant powering applications [12]. The results show a significant improvement in efficiency. Figure 3.5 illustrates the equivalent electrical model of the 4-coil resonant inductive link. It is proposed that the low coupling coefficient and the low-quality factor of the coils in the 2-coil link can be compensated by the introduced two high-quality factor coils between them [21].

The nature of implanted systems imposes the usage of smaller coils, which limits the coupling between the external and internal inductors. Assuming the strong or medium coupling between the inductors on the same side and weak coupling across the skin except k_{23} enables to neglect the coupling coefficients k_{14}, k_{13}, and k_{24}. Accordingly, the PTE of the 4-coil inductive link can be expressed as

$$\eta = \frac{\left(k_{12}^2 Q_1 Q_2\right)\left(k_{23}^2 Q_2 Q_3\right)\left(k_{34}^2 Q_3 Q_4\right)}{\left[\left(1+k_{12}^2 Q_1 Q_2\right)\left(1+k_{34}^2 Q_3 Q_4\right)+k_{23}^2 Q_2 Q_3\right]\left[1+k_{23}^2 Q_2 Q_3+k_{34}^2 Q_3 Q_4\right]}.$$
$$(3.3)$$

Further assumptions simplify the efficiency expression while introducing new design constraints. Accepting the moderate coupling between L_1 and L_2 (k_{12}), and L_3 and L_4 (k_{34}), as well as the high-quality factor of Q_2 and Q_3, provide the following assumptions:

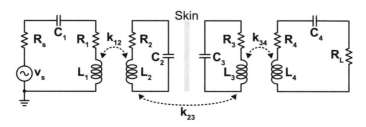

Fig. 3.5 Equivalent electrical model of the 4-coil resonant inductive link

$$k_{12}^2 Q_1 Q_2 \gg 1 \tag{3.4}$$

$$k_{34}^2 Q_3 Q_4 \gg 1 \tag{3.5}$$

$$\left(k_{12}^2 Q_1 Q_2\right)\left(k_{34}^2 Q_3 Q_4\right) \gg k_{23}^2 Q_2 Q_3 \tag{3.6}$$

$$1 + k_{23}^2 Q_2 Q_3 \gg k_{34}^2 Q_3 Q_4 \tag{3.7}$$

and result in the simplest form of efficiency for 4-coil inductive link.

$$\eta \cong \frac{k_{23}^2 Q_2 Q_3}{1 + k_{23}^2 Q_2 Q_3}. \tag{3.8}$$

The approximated efficiency relation for the 4-coil inductive link is similar to the efficiency of the 2-coil structure except for the source and load resistances. As a result, the introduced coils, L_2 and L_3, transform any arbitrary load impedance to the optimal impedance at the input of the inductive link and efficiency does not change significantly with load power. Likewise, a 3-coil inductive link is presented [22] for higher power loads, but a low power load application favors 4-coil structure to provide high PTE.

In order to take advantage of high PTE and power load tolerance, a 4-coil inductive link is implemented for epilepsy monitoring. The resonance frequency of the coils is selected as 8 MHz in order to minimize the losses in tissue [23]. For determining the geometric parameters of the coils, an optimization algorithm is used with the maximum PTE as a goal function [24]. The dimension of the Burr hole determines the maximum outer diameter and the fabrication capabilities of printed circuit boards limit the metal width and spacing. Resulting coil dimensions and measured electrical characteristics are given in Table 3.5. The designed inductive link is manufactured on a 0.8 mm thickness FR4 board with 18 μm copper thickness. Figure 3.6 shows the fabricated coils.

Table 3.5 Geometric and electrical parameters of designed coils

	Source coil (L_1)	Primary coil (L_2)	Secondary coil (L_3)	Load coil (L_4)
Inner diameter (mm)	13	29	6.9	1.9
Outer diameter (mm)	24	43	13.5	5.5
Width and spacing (mm)	0.5	1	0.3	0.2
Number of turns	6	4	6	5
Inductance (μH)	1.10	1.14	0.59	0.14
Quality factor	48	64	30	21
Tuning capacitance (pF)	330	326	616	2320

Fig. 3.6 Images of the fabricated coils: L_1 (left-inner), L_2 (left-outer), L_3 (right-outer), and L_4 (right-inner)

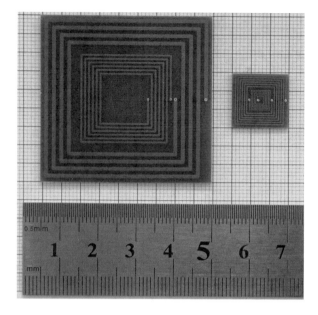

The fabricated inductive link was characterized with 10 mW load power at a distance of 10 mm. Comparing to the 2-coil solution with the same source and load coil geometries, the PTE is increased from 37 to 55%. Despite the difficulty of resonating 4 coils at the same frequency, improvement in PTE and better performance with different loads justify the 4-coil inductive link solution for the desired application.

3.4.2 Rectifier

The induced AC voltage on the implanted coil requires to be converted by the rectifier to a DC voltage, which is going to be the supply of the monitoring system. The performance of the WPT receiver strongly affects the PTE and available power to the implanted electronics. Diodes are the simplest semiconductor devices which allow the current flow only in one direction. This property of diodes is used in half-wave and full-wave rectifications shown in Fig. 3.7a, b.

A diode has a forward voltage drop and minimization of it provides a higher voltage conversion ratio. A Schottky diode formed by a junction between a semiconductor and metal provides a low voltage drop and it is widely used in discrete solutions. However, creating a Schottky diode in the CMOS process requires additional masks and creates extra manufacturing costs. Alternatively, integrated rectifiers use diode connected transistors, as illustrated in Fig. 3.7d, e. It has to be noted that the bulk of the diode connected transistor is connected to

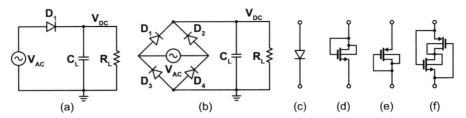

Fig. 3.7 Schematics of (**a**) the half-wave rectifier, (**b**) the full-wave rectifier, (**c**) the diode, (**d**) diode connected NMOS, (**e**) diode connected PMOS, and (**f**) composite CMOS diode

its drain terminal as opposed to the conventional applications in analog circuits. The leading cause of this type of connection is to achieve the same polarity of the parasitic diode as the transistor diode to reduce the leakage current when it is reverse biased.

Further improvement to minimize the reverse leakage current of the diode connected transistor is to use the composite CMOS diode [25], as shown in Fig. 3.7f. In the forward-biased mode, the composite CMOS diode behaves as a series connection of two diodes. When the composite diode is reversely biased, the V_{DS} voltages are halved compared to the diode connected transistor because the voltage across the two terminals can be approximately expressed as

$$|V_{gs\text{NMOS}}| = |V_{gs\text{PMOS}}| = |V_{ds\text{NMOS}}| + |V_{ds\text{PMOS}}|. \tag{3.9}$$

The reduction of the V_{DS} voltage in reverse-biased mode lowers the leakage current. It is demonstrated a reduction of leakage currents by four orders of magnitude compared to standard MOS diode implementation [25]. However, the benefit of the composite diode comes with the cost of doubling the voltage drop in forward-biased operation. Chen et al. presented a full-wave rectifier based on the composite diode structure in parallel with a controlled switch for reducing the voltage drop and ended up an active rectifier [26]. In the proposed architecture, the comparator detects the moment when the input voltage exceeds the output voltage level and turns on the switch to enable the conduction. This method provides low forward voltage compared to passive rectification, however, consumes additional power for the comparator. It is only beneficial to operate with an active rectifier when the gain of voltage drop reduction surpasses the losses because of switching decision and driver. The power consumption of the comparator can be considered as constant and does not depend on the input power. Therefore, the active rectifier performs better efficiency for power levels higher than the mW regime, while the passive rectification offers a good conversion for ultra-low power applications.

Considering the 10 mW power transmission is needed to the neural monitoring system, a half-wave rectifier based on a composite diode in parallel with a comparator-controlled switch is adopted for the conversion of induced AC voltage to a DC voltage. The schematic of the designed active half-wave rectifier is presented

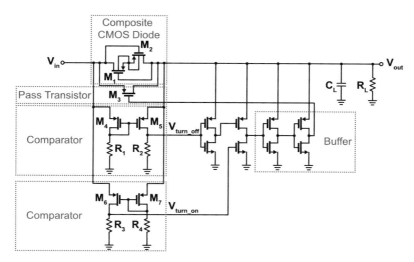

Fig. 3.8 Schematic of the active half-wave rectifier

in Fig. 3.8. Two PMOS transistors are used to create the composite diode since NMOS transistors particularly require a twin-well fabrication process. A PMOS pass transistor placed in the same n-well with the composite diode is used as a switching element. The comparator constituted by the transistors M_4–M_7 decides to turn on or turn off the switch. There is a need for a buffer stage between the comparator and the switch to be able to drive the low resistive PMOS pass transistor.

The comparison of input and output voltages relies on the parallel operation of two comparators constituted by M_4–M_5 and M_6–M_7 PMOS transistors. When the output voltage is high enough, the diode connected M_7 transistor turns on and the low current passing through the resistance R_4 creates a potential at the gate, V_{g7}.

$$V_{g7} = V_{out} - |V_{gs7}| \tag{3.10}$$

When the input potential is higher than V_{g7} by an amount of threshold voltage, V_{th6}, the transistor M_3 starts to conduct and creates a turn-on voltage. The condition for the high state of turn-on voltage can be expressed by

$$V_{in} > V_{out} - |V_{gs7}| + |V_{th6}|. \tag{3.11}$$

Proper choice of component values for a particular amount of output voltage yields to equality of V_{gs7} and V_{th6} and the requirement for the generation of turn-on voltage becomes

$$V_{in} > V_{out}. \tag{3.12}$$

Using the same technique, a turn-off voltage is generated when the output voltage overshoots the input voltage. The high state of turn-on voltage creates a low voltage to allow the current flow on the PMOS pass transistor. On the contrary, a high state of turn-off voltage generates a high control voltage to eliminate the reverse current from output to input. The switching time of the pass transistor is crucial and any latency creates a decrease in efficiency. The two inverter buffer stage has a significant role in reducing the switching time of the pass transistor.

The proposed active half-wave rectifier was designed in UMC 180 nm process technology. The rectifier occupies an area of $110\,\mu m \times 250\,\mu m$. The core voltage of the technology is 1.8 V. Therefore, the targeted output voltage of 2 V for the regulator is decided to take into account the drop voltage caused by the linear regulator. The selected load capacitance is 100 nF. Figure 3.9 shows the V_{in}, V_{out}, and V_{g3} voltages as well as the input current, I_{in}, during the rectification operation for 2 V output voltage and 10 mW load. There are delays of 4.7 and 5 ns in the switching of the pass transistor. Late turning on of the pass transistor limits the delivered current to the output and delayed turn-off command creates a leakage

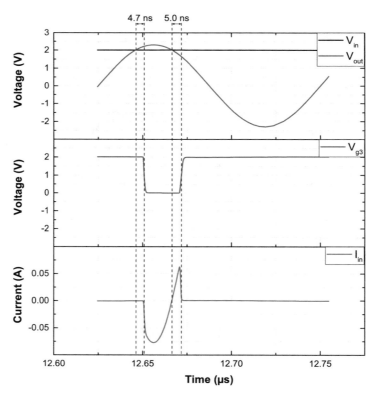

Fig. 3.9 Waveforms of input, output, and pass transistor gate voltages as well as the input current for a duration of one period of the input wave

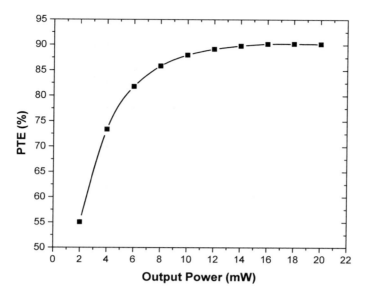

Fig. 3.10 PTE of the active rectifier for different output power levels

current. The latencies cause a deviation from the ideal performance of rectification and results with 88% of PTE.

Figure 3.10 shows the PTE for various load conditions while keeping the output voltage the same at 2 V. For the low level of load power, active rectifier performs poor PTE because the losses in the comparator and buffer stages are comparable with the output power. However, for the higher level of output power, the power delivered to the load becomes dominant and an active rectifier shows good efficiency. The plot confirms that the active rectifier is preferable for 10 mW and higher output power applications rather than the ultra-low power systems.

The micrograph of the fabricated rectifier is shown in Fig. 3.11. An efficiency of 82% is measured for the desired operation. Table 3.6 summarizes the performance of the proposed rectifier with state-of-the-art designs.

3.4.2.1 Improvements for the Rectifier

The latencies limit the performance of the designed rectifier in the switching of the pass transistor after the comparison of input and output voltages. The delay created by the buffer stage is inevitable even if the comparator detects the switching moments in a very short duration. A possible solution to overcome this drawback in the active rectification method is to use an offset voltage in one of the inputs of the comparator to start the switching action in advanced [27]. Instead of comparison between the voltages at the input and the output of the rectifier, the input voltage can be compared with a potential which is smaller than the output voltage in a certain

Fig. 3.11 Micrograph of the
active half-wave rectifier

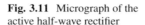
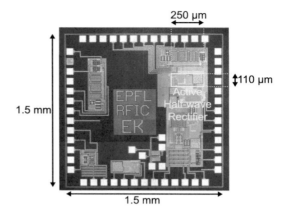

Table 3.6 Comparison of the designed half-wave active rectifier with similar works in the literature

	Process technology (nm)	Operation frequency (MHz)	Load capacitance (nF)	Output voltage (V)	Output power (mW)	Measured PTE (%)
[27]	500	13.56	10,000	2.5–3.9	30.42	68–80.2
[28]	180	10	0.2	0.2–2	2	37–80
[29]	350	13.56	1.5	1.19–3.52	24.8	82.2–90.1
[30]	180	13.56	10,000	1.33	1.7	81.9
[31]	180	13.56	0.2	0.39–2.92	13.72	87
This work	180	8	100	2	10	82.0

amount so that the turning on time of the switch matches well with the time when input voltage and output voltage are equal. Likewise, the turning off time of the pass transistor can be adjusted. However, this method requires additional circuits and attention to the decision mechanism. The new conditions for turning on (3.13) and turning off (3.14) the switch can be represented as

$$V_{in} > V_{out} - V_{offset1} \qquad (3.13)$$

and

$$V_{in} < V_{out} + V_{offset2}. \qquad (3.14)$$

In the case of there is no offset voltage in the comparator, the turn-on and turn-off commands are generated when the input voltage is higher or lower than the output voltage, respectively. The waveforms for input and output voltages, as well as the ON and OFF states of the pass transistor are illustrated in Fig. 3.12a. As it is demonstrated, there is no condition that can produce both of the signals at the

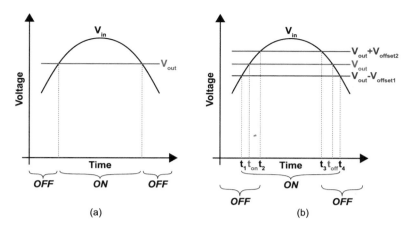

Fig. 3.12 Timing diagram of the PMOS pass transistor states for (**a**) without and (**b**) with offset voltages in the comparator input

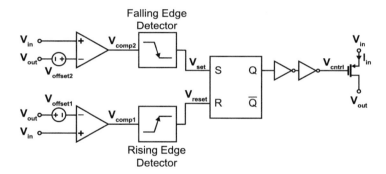

Fig. 3.13 Proposed structure for improved performance of active rectifier

same moment except the crossing points when the input voltage equals the output voltage. On the other hand, as a result of applied offset voltages to the comparators and imposing the conditions in Eqs. 3.13 and 3.14, Fig. 3.12 illustrates that there are two overlapping regions where the pass transistor is supposed to be in ON and OFF state at the same time. For the intended operation, OFF state before the ON state has been ended up at the t_1 time and ON condition needs to stop at t_3. The correct detection of t_1 and t_3 times requires not only the voltage level but also the slope information of the input signal. The proposed structure in Fig. 3.13 can achieve the interested operation and eliminates the overlying decisions by using the rising edge and falling edge detectors.

Instead of relying on the continuous-time comparison and controlling the pass transistor, the crossing points of the input voltage and comparison levels are detected. The correct operation of the rectifier requires to create a turn-on signal

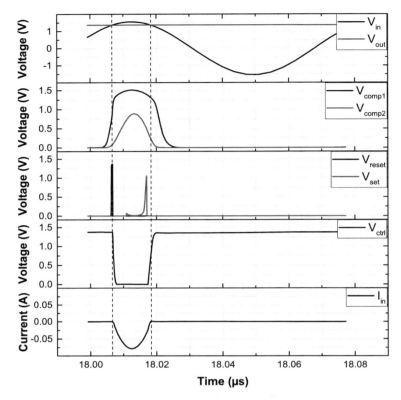

Fig. 3.14 Post-layout simulation waveforms in the proposed structure for improved performance of active rectifier for a duration of one period of the input signal

at time t_1 and turn-off signal at t_3. The moment t_1 is captured by comparing the V_{in} with $V_{out} - V_{offset1}$ voltage and detecting the rising edge of the comparison result. Just after the t_1 time, the rise edge detector creates a short pulse and resets the output of the set-reset (SR) latch. Then pass transistor starts to allow the conduction of current from the input port to the output port. The same principle can be applied to detect the moment when V_{in} becomes smaller than $V_{out} + V_{offset2}$. A falling edge detector follows the output of the comparator and a short pulse is created at time t_3. The created pulse is connected to the set port of the SR latch. The generated high voltage at the output of the SR latch is buffered and arrives in the gate of the pass transistor at t_{off} to stop the conduction. The visual representation of the operation is illustrated in Fig. 3.14.

The improved version of the proposed active half-wave rectifier is designed in TSMC 65 nm technology. The fabricated chips are not delivered and not characterized at the moment of writing the book. Accordingly, post-layout simulations are shown for performance evaluation of the designed block. The output power

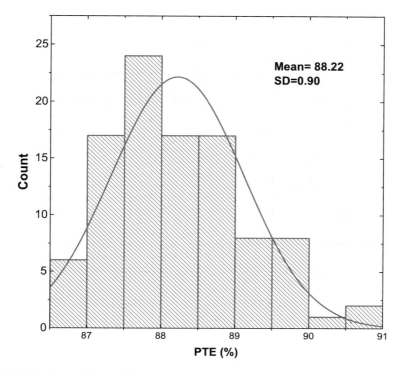

Fig. 3.15 Post-layout Monte Carlo simulation of the PTE with a hundred runs

specification of the rectifier is kept the same at 10 mW while reducing the output voltage from 2 to 1.3 V. The operating frequency is selected as 13.56 Mhz in industrial, scientific, and medical (ISM) radio bands. The design occupies an 85 μm × 95 μm area.

Monte Carlo simulation with a hundred runs is conducted for post-layout design to ensure the performance of the circuit. The current flow is maximized in forward-biased mode and the leakage current from output to input minimized for reverse-biased configuration. Figure 3.15 illustrates the power transfer efficiency histogram for a hundred runs. The mean of the PTE and standard deviation (SD) are 88.22% and 0.90%, respectively. Compared to the measured PTE of the previously designed active rectifier, there is an improvement in the PTE by reducing the leakage and maximizing the forward currents for the same output powers.

The offset voltages in the improved version are optimized for the output voltage of 1.3 V. However, for operations which require different output voltages, it is necessary to readjust them. To the best of action, a feedback loop may control the offset voltages with the goal of maximizing the output voltage. Moreover, the process variations can be eliminated with a feedback loop.

3.4.3 Regulator

The designed implantable circuits are optimized for constant DC supply voltages. The maximum performance and interested application require stable and ripple-free supply voltage. However, the proposed methodology for powering the integrated circuits relies on the magnetically induced alternating current and the output of the rectifier contains small ripples at the transmission frequency. Furthermore, there is a fluctuation in the received power since powering coils are not mechanically fixed so that distance and the coupling between the powering coils are not constant. The lateral and angular misalignment between the coils can cause considerable variations in the power transmission. Besides, the power consumption of the system is changing depending on the operation mode and instantaneous changes in current drawn from supply generate fluctuations. In order to cover the mentioned issues, a regulator block following the rectifier is necessary. Similar to the inductive link and rectifier block, the regulator is a part of the supply generation chain, and its efficiency directly affects the system efficiency. To achieve a low loss in the regulation stage, a low drop-out (LDO) voltage regulator is used in the proposed system.

The basic LDO architecture is shown in Fig. 3.16. The error amplifier, reference generator, and power transistor are connected to the output of the rectifier. A feedback loop is constituted by the error amplifier, power transistor, R_1 and R_2 resistors. The error amplifier determines the gate voltage of the power transistor so that the divided output voltage becomes the same as the generated reference voltage.

The power transistor can be chosen as NMOS or PMOS, depending on the application. The configuration with the NMOS transistor provides better regulation because its low output impedance creates a small RC time constant at the output. Also, the higher mobility of the NMOS transistor reduces the required area. However, the selection of the NMOS transistor requires a step-up converter to be able to drive the power transistor when the voltage difference between the input and output of the rectifier is smaller than the V_{gs}. On the contrary, for the case of the PMOS transistor, the requirement for a low drop-out application is to lower the voltage at the gate of the transistor. Accordingly, the regulator with a PMOS power transistor provides low voltage drop and high power efficiency without increasing

Fig. 3.16 Schematic of an
LDO regulator

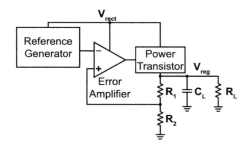

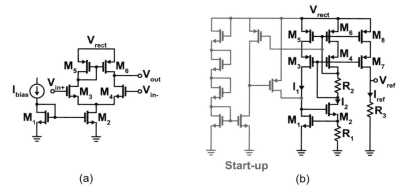

Fig. 3.17 Schematic of (**a**) an OTA and (**b**) a reference generator

the complexity of the circuit. In aimed regulation application, PMOS is selected as the type of power transistor.

The precision and the response time of the regulation are determined by, respectively, the gain and bandwidth of the error amplifier. The removal of the ripples at the remote powering frequency requires an amplifier with large bandwidth. An operational transconductance amplifier (OTA) has been decided to be used as the error amplifier since it offers higher bandwidth compared to operational amplifiers (OPAMP). It has been shown that the single-stage OTA provides an order of magnitude better power supply rejection than the two-stage Miller OTA [32]. The preferred single-stage OTA is presented in Fig. 3.17a.

The LDO needs a reference voltage at the input of the error amplifier to compare the divided version of the regulated voltage. The reference generator is supplied by the unregulated output of the rectifier, which is not stable. Hence, the reference voltage is supposed to be independent of the supply voltage. A self-biased supply independent current generator has been adopted for the core of the reference generator. Figure 3.17b shows the cascoded solution to cope with the channel length modulation on the current mirror M_3–M_4 and the start-up circuit. If the I_1 and I_2 currents are kept the same with the two current mirrors M_3–M_4 and M_5–M_6, then the V_{gs1} can be expressed by two equations.

$$V_{gs1} = R_1 I_1 \tag{3.15}$$

$$V_{gs1} = V_{th1} + \sqrt{\frac{I_1}{\frac{\mu C_{ox}}{2}\left(\frac{W_1}{L_1}\right)}} \tag{3.16}$$

for $V_{gs1} > V_{th1}$. Solving these two equations gives two operating points. One of the solutions forces I_1 and V_{gs1} to be zero. In order to this condition, the start-up circuit presented in Fig. 3.17b is needed. The other operating point defines the

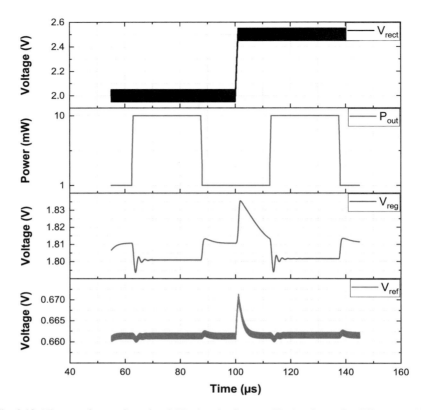

Fig. 3.18 The waveforms of regulated (V_{reg}) and reference (V_{ref}) voltages for different rectified voltage (V_{rect}) and output power (P_{out}) levels

supply independent current I_1 in terms of V_{gs1}. The current I_1 can be mirrored with the transistors M_3–M_4 and the resistor R_3 creates the reference voltage, V_{ref}. The reference voltage can be used not only in the LDO regulator but also in the other circuits in the implanted system. Moreover, the need for the bias currents of the circuits such as OTA can be fulfilled by mirroring the current I_{ref}.

The proposed LDO regulator was designed in UMC 180 nm process technology. The regulator occupies an area of $125 \, \mu m \times 200 \, \mu m$. Figure 3.18 shows the regulation of the output voltage for the changes in the load and the unregulated voltage as well as the stability of the generated reference voltage. The increase in the rectifier output from 2 to 2.5 V raises the regulated voltage by only 35 mV. Likewise, the increasing output power by ten times reduces the output voltage just 20 mV for a short time and the LDO quickly reflects and adjusts the output voltage level to 1.8 V.

The micrograph of the fabricated regulator is shown in Fig. 3.19. An efficiency of 78% is measured for 10 mW regulated output when the unregulated input is 2 V and load capacitance is 100 nF.

Fig. 3.19 Micrograph of the
LDO regulator

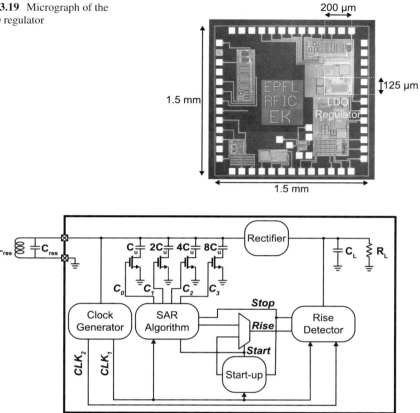

Fig. 3.20 Proposed structure for automatic resonance tuning for an inductively coupled remote
powering system

3.4.4 Automatic Resonance Tuning

The power transfer efficiency through magnetically coupled coils is maximized
when the coils resonate at precisely the same frequency. However, this resonance
frequency is highly sensitive to the process variations and requires a significant
amount of effort for manual trimming in order to reach the same frequency. Even
worst case, manual trimming of the resonant circuit cannot be a solution for the
shift in the resonance frequency caused by packaging and the nearby tissue effects
in implanted biomedical applications. The frequency of the external base station
can be fixed to a specific value, but there is no more control over the implanted
system. Therefore, there is a need for an automatic resonance tuning mechanism that
maximizes the received power. Figure 3.20 shows the developed system inspired by
the work presented in [33], which controls the bank of capacitors to maximize the
received regulated voltage for a constant load.

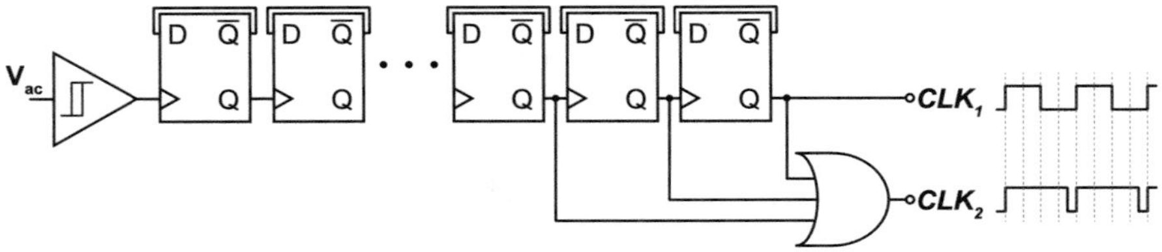

Fig. 3.21 Schematic of a clock generation in inductively coupled remote powering system

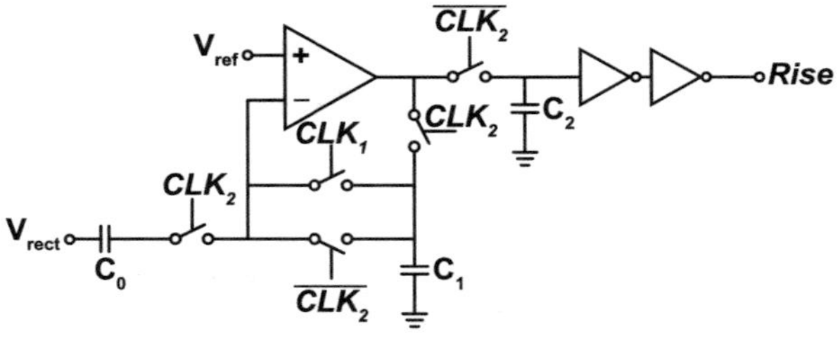

Fig. 3.22 Schematic of the rise detector

The automatic resonance tuning system is designed by the building blocks of a clock generator, rise detector, start-up, successive approximation register (SAR) algorithm, and a bank of capacitors. The rise detector, start-up, and SAR algorithm blocks work in synchronously and require clock signals. The two clocks, CLK_1 and CLK_2, are generated by using a Schmitt Trigger, a chain of D-flip-flops and an OR-gate, as shown in Fig. 3.21. Schmitt trigger uses the induced AC voltage as the input and creates a square wave at the same frequency with remote powering. Although the generated square wave does not have a 50% duty cycle because of the hysteresis in the Schmitt trigger, the duty cycle is corrected in the following stages of D-flip-flops (DFF). There are nine stages of DFF to divide the powering frequency by 512. Generated CLK_1 clock is used by all the blocks in the automatic resonance tuning mechanism. Besides, the rise detector block needs one more block at the same frequency with CLK_1 but higher duty cycle. A three-input OR logic gate uses the output of the last three stages of DFF and creates CLK_2. Similar to the case of the reference generator in the LDO regulator, the clock generator block can be used in other applications when there is a need for a clock with a frequency smaller or equal to powering frequency.

The rise detector illustrated in Fig. 3.22 uses the generated two clocks and check whether there is an increase in the rectified voltage or not in every period of the clocks. When both of the clocks are at a high state, the current state of the rectifier voltage is stored in capacitor C_1. When the SAR algorithm is operating, a short duration after the CLK_1 gets low state, one of the digitally controlled capacitors turns on, and the new voltage at the output of the rectifier is compared with the previously stored voltage. If there is an increase in the voltage, the rise detector generated a high state when CLK_2 gets lower. SAR algorithm takes the output of the rise detector and decides to keep the control bit of the turned on capacitor high. In the case of there is no detected increase, then the capacitor is deactivated by the SAR algorithm.

The tuning algorithm works at each time that the system is powered. When a remote powering becomes active, the capacitor at the output of the rectifier is charged slowly. There is a need to distinguish the increase at the output is because of the nature of the powering up the system or because of the SAR algorithm. Therefore, the proposed system waits enough to start the SAR algorithm before settling down the rectified voltage. The output of the start-up block is low by default before the rectifier output becomes stable. The output of the rise detector is directed to the start-up block by the demultiplexer. The *start* command gets active when there is no increase in V_{rect} for four consecutive measurements. The high state of the *start* command connects the rise detector to the SAR algorithm block and initiates the tuning mechanism. When the SAR algorithm finishes, a *stop* signal is generated, and the blocks used in the automatic tuning is turned off to save power.

The proposed system was designed in UMC 180 nm process technology. The fabricated chips are not delivered and not characterized at the moment of writing the book. Accordingly, post-layout simulations are shown for performance evaluation of the designed block. The automatic resonance tuning occupies an area of $600\,\mu\text{m} \times 330\,\mu\text{m}$. The operation frequency is chosen 13.56 MHz, and the equivalent inductance of the remote powering coil is selected as $L_{res} = 4.6\,\mu\text{H}$. The unit capacitance of $C_u = 480\,\text{fF}$ is used in the digitally controlled capacitor bank.

Figure 3.23 shows the rectified voltage, V_{rect}, and the control bits of the capacitors during the tuning algorithm in the post-layout simulation when the external resonance capacitor value, C_{res} is chosen as 24.1 pF. The rectified voltage is increased for the high values of C_3 and C_2 while activation of C_1 and C_0 creates a drop in V_{rect}. The algorithm provides the value of the equivalent capacitance value of

$$C_{res} + 12C_u = 29.86\,\text{pF} \tag{3.17}$$

which is very close to the ideal capacitance of 29.95 pF. The tuning process takes about $265\,\mu\text{s}$ and consumes only $2.3\,\mu\text{W}$. Figure 3.24 shows the rectified voltage for variations in LC product with and without tuning algorithm. Considering that the tolerable voltage drop in V_{rect} is 5%, the automatic resonance tuning provides tolerance of $\pm 11.0\%$ variation in LC product while it is limited to $\pm 2.9\%$ without the proposed method.

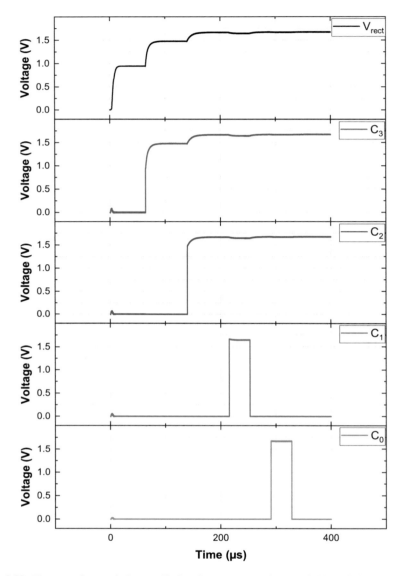

Fig. 3.23 The waveform of the rectified voltage V_{rect} and control bits of the capacitors, $C_3 C_2 C_1 C_0$, during the tuning mechanism

3.4.5 *Power Feedback Generation*

The automatic resonance tuning mechanism offers to maximize the power transfer efficiency in remote powering. However, there is a missing information about whether the delivered power is suitable for the optimum operation. There are mainly

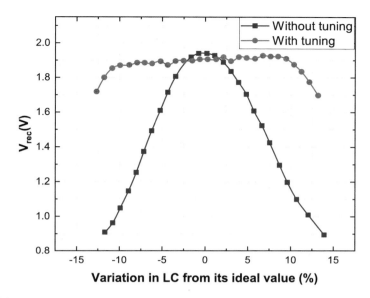

Fig. 3.24 Post-layout simulation of the rectified voltage V_{rect} with and without tuning mechanism as a function of LC shift

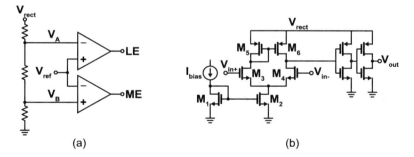

Fig. 3.25 Schematic of (**a**) power feedback generation block and (**b**) comparator used in power feedback

three important reasons to watch the delivered power. Firstly, there should be enough power to be able to operate the implanted system as expected. Secondly, an excessive amount of induced voltage over the implanted coils create a damage in the rectifier and regulator. The last and the most important aspect of delivering more than expected power to the implanted system creates a waste of energy and temperature increase in the tissue. Accordingly, there is a clear demand for a feedback mechanism to monitor and adjust the power delivered to the monitoring system. A very simple and effective system can be developed by using the power level detector in Fig. 3.25.

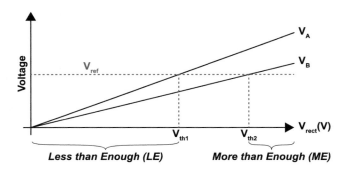

Fig. 3.26 Voltage waveforms and determination of threshold values in power feedback generation

As in the case of automatic resonance tuning algorithm, monitoring rectified voltage provides enough information about the received power. The proposed system in Fig. 3.25a divides the voltage at the output of the rectifier and generates V_A and V_B voltages to make them comparable with the reference voltage. The generated supply independent reference voltage, V_{ref}, for the low drop-out voltage regulation is used also in power feedback generation. The change in V_A and V_B voltages with the change in V_{rect} is illustrated in Fig. 3.26.

The intersection points of generated V_A and V_B voltages with V_{ref} voltage define the desired minimum and maximum levels of the rectified voltage. At the end of comparison using the presented comparator in Fig. 3.25b the two output bits, namely less than enough (LE) and more than enough (ME), provide the necessary information. When V_A is smaller than V_{ref}, the top comparator in Fig. 3.25a makes LE bit high. In other words, LE bit shows whether the V_{rect} is higher or lower than V_{th1} voltage. Likewise, the bottom comparator in Fig. 3.25a sets ME bit high when V_B exceeds the reference voltage. The high state of ME bit indicates that V_{rect} is higher than V_{th2} and shows that there is abundant power than what is needed. The two threshold values, V_{th1} and V_{th2}, can be fixed to the desired values by proper choice of the voltage divider resistors. ME and LE bits are delivered to the external base station thanks to uplink data communication, and the external coil driver is readjusted if there is a need.

The proposed system was designed in UMC 180 nm process technology. The fabricated chips are not delivered and not characterized at the moment of writing the book. Accordingly, post-layout simulations are shown for performance evaluation of the designed block. The power feedback generation occupies an area of $260\,\mu$m $\times$ $100\,\mu$m. Figure 3.27 shows the behavior of LE and ME outputs for various input voltages. The optimum driving current of the external coil has to be chosen to make LE and ME low at the same. The values of the components in resistive divider in Fig. 3.25a are selected such that the active rectifier performs the optimum operation ($V_{rect} = 2$ V).

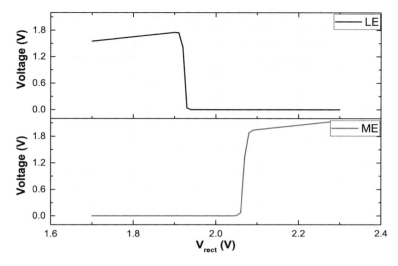

Fig. 3.27 The waveform of the less than enough (LE) and more than enough (ME) signals for different values of rectified voltage, V_{rect}

3.5 Summary

This chapter explains how wireless power transfer is established for the implanted neural monitoring system. The approach for powering the implanted system up and its applicability to the desired operation are analyzed. The hybrid solution of battery and remote powering selection as the power supply is justified. A magnetically coupled remote powering is selected for charging of the secondary battery considering the specifications in the project. The choice of 4-coil inductive coupling is explained. The active half-wave rectifier and LDO regulator blocks are presented for the generation of stable and ripple-free supply voltage for the implanted system. Additionally, the automatic resonance tuning mechanism is developed to maximize the PTE. Furthermore, the power feedback generation structure is designed to make sure that the delivered power is enough for the operation. The characterizations of the presented circuits are done in measurement or post-layout simulation level depending on the availability of the fabricated chips.

References

1. IEEE (2006) IEEE standard for safety levels with respect to human exposure to radio frequency electromagnetic fields, 3 kHz to 300 GHz. IEEE Std C951-2005 (revision of IEEE Std C951-1991). IEEE, New York, pp 1–238
2. Yilmaz G (2014) Wireless power transfer and data communication for intracranial neural implants case study. École Polytechnique Fédérale de Lausanne (EPFL), Lausanne

3. Antonioli G, Baggioni F, Consiglio F et al (1973) Stinulatore cardiaco impiantabile con nuova battaria a stato solido al litio. Minerva Med 64:2298–2305
4. Mond HG, Proclemer A (2011) The 11th World survey of cardiac pacing and implantable Cardioverter-Defibrillators: calendar year 2009–a world society of arrhythmia's project. Pacing Clin Electrophysiol 34:1013–1027
5. Roundy S, Wright PK, Rabaey JM (2004) Energy scavenging for wireless sensor networks: with special focus on vibrations. Springer, Berlin
6. Kwon D, Rincon-Mora GA (2010) A 2-μm BiCMOS rectifier-free AC–DC piezoelectric energy harvester-charger IC. IEEE Trans Biomed Circuits Syst 4:400–409
7. Zhang Y, Zhang F, Shakhsheer Y et al (2013) A batteryless 19 μW MICS/ISM-band energy harvesting body sensor node SoC for ExG applications. IEEE J. Solid State Circuits 48:199–213
8. Ayazian S, Hassibi A (2011) Delivering optical power to subcutaneous implanted devices. In: 2011 Annual international conference of the IEEE engineering in medicine and biology society, pp 2874–2877
9. Goto K, Nakagawa T, Nakamura O, Kawata S (2001) An implantable power supply with an optically rechargeable lithium battery. IEEE Trans Biomed Eng 48:830–833
10. Chow EY, Yang C, Ouyang Y et al (2011) Wireless powering and the study of RF propagation through ocular tissue for development of implantable sensors. IEEE Trans Antennas Propag 59:2379–2387
11. Ho JS, Kim S, Poon ASY (2013) Midfield wireless powering for implantable systems. Proc IEEE 101:1369–1378
12. Yilmaz G, Atasoy O, Dehollain C (2013) Wireless energy and data transfer for in-vivo epileptic focus localization. IEEE Sensors J 13:4172–4179
13. Sauer C, Stanacevic M, Cauwenberghs G, Thakor N (2005) Power harvesting and telemetry in CMOS for implanted devices. IEEE Trans Circuits Syst Regul Pap 52:2605–2613
14. Catrysse M, Hermans B, Puers R (2004) An inductive power system with integrated bi-directional data-transmission. Sens Actuators, A 115:221–229
15. Mazzilli F, Thoppay PE, Praplan V, Dehollain C (2012) Ultrasound energy harvesting system for deep implanted-medical-devices (IMDs). In: 2012 IEEE international symposium on circuits and systems, pp 2865–2868
16. Mathieson K, Loudin J, Goetz G, et al (2012) Photovoltaic retinal prosthesis with high pixel density. Nat Photonics 6:391–397
17. Lee SB, Lee B, Kiani M et al (2016) An inductively-powered wireless neural recording system with a charge sampling analog front-end. IEEE Sensors J 16:475–484
18. US Food and Drug Administration (1997) Information for manufacturers seeking marketing clearance of diagnostic ultrasound systems and transducers. Center for Devices and Radiological Health, US Food and Drug Administration, Rockville
19. Baker MW, Sarpeshkar R (2007) Feedback analysis and design of RF power links for low-power bionic systems. IEEE Trans Biomed Circuits Syst 1:28–38
20. Kurs A, Karalis A, Moffatt R et al (2007) Wireless power transfer via strongly coupled magnetic resonances. Science 317:83–86
21. RamRakhyani AK, Mirabbasi S, Chiao M (2011) Design and optimization of resonance-based efficient wireless power delivery systems for biomedical implants. IEEE Trans Biomed Circuits Syst 5:48–63
22. Kiani M, Jow U, Ghovanloo M (2011) Design and optimization of a 3-coil inductive link for efficient wireless power transmission. IEEE Trans Biomed Circuits Syst 5:579–591
23. Vaillancourt P, Djemouai A, Harvey JF, Sawan M (1997) EM radiation behavior upon biological tissues in a radio-frequency power transfer link for a cortical visual implant. In: Proceedings of the 19th annual international conference of the IEEE engineering in medicine and biology society. "Magnificent milestones and emerging opportunities in medical engineering" (Cat. No.97CH36136), vol 6, pp 2499–2502
24. Silay KM (2012) Remotely powered wireless cortical implants for brain-machine interfaces. École Polytechnique Fédérale de Lausanne (EPFL), Lausanne

25. Levacq D, Liber C, Dessard V, Flandre D (2004) Composite ULP diode fabrication, modelling and applications in multi-Vth FD SOI CMOS technology. Solid State Electron 48:1017–1025
26. Chen C-L, Chen K-H, Liu S-I (2007) Efficiency-enhanced CMOS rectifier for wireless telemetry. Electron Lett 43:976–978
27. Lee H, Ghovanloo M (2011) An integrated power-efficient active rectifier with offset-controlled high speed comparators for inductively powered applications. IEEE Trans Circuits Syst Regul Pap 58:1749–1760
28. Hashemi SS, Sawan M, Savaria Y (2012) A high-efficiency low-voltage CMOS rectifier for harvesting energy in implantable devices. IEEE Trans Biomed Circuits Syst 6:326–335
29. Lu Y, Ki W (2014) A 13.56 MHz CMOS active rectifier with switched-offset and compensated biasing for biomedical wireless power transfer systems. IEEE Trans Biomed Circuits Syst 8:334–344
30. Cha H, Park W, Je M (2012) A CMOS rectifier with a cross-coupled latched comparator for wireless power transfer in biomedical applications. IEEE Trans Circuits Syst Express Briefs 59:409–413
31. Khan SR, Choi G (2017) High-efficiency CMOS rectifier with minimized leakage and threshold cancellation features for low power bio-implants. Microelectron J 66:67–75
32. Steyaert MSJ, Sansen WMC (1990) Power supply rejection ratio in operational transconductance amplifiers. IEEE Trans Circuits Syst 37:1077–1084
33. Gosselin P, Puddu R, Carreira A et al (2017) A CMOS automatic tuning system to maximize remote powering efficiency. In: 2017 IEEE international symposium on circuits and systems (ISCAS), pp 1–4

Chapter 4
Wireless Data Communication

4.1 Introduction

Observing the complex neural dynamics of the brain is desired by research communities for discovering neurophysiological behaviors. Recording of a high number of channels is needed in brain–machine interface (BMI) systems since the nervous system can only be understood by simultaneous recording of large neuronal ensembles. For example, according to the recent estimations, 100,000 neurons need to be monitored in order to explain the full-body movement [1]. Monitoring and analyzing these neural activities play a significant role in the treatment of neurological disorders such as traumatic brain injury, Parkinson's disease, or epilepsy. In the content of this book, the focus is mainly on implanted intracranial recordings for epilepsy; however, the scope of the study can be extended to various monitoring applications.

The transition from a conventional intracranial EEG monitoring method to an implantable iEEG system for presurgical analysis of epilepsy disease brings some challenges. One of the significant difficulties to cope with is to eliminate the cables which carry the measurements of the neural activity. The implanted ADC digitalizes the acquired analog measurements by the microelectrode arrays. There is a need for an uplink communication to transmit the digitalized data to the external base station for the analysis before the surgery. Moreover, the uplink data transmission is also necessary to get the status of the implanted device. For instance, the information about wirelessly transferred power level can be sent to the external unit, and the appropriate action is taken depending on the situation. In addition to the uplink communication, there is a requirement of downlink data communication from the external station to the implanted system to configure various parameters such as an active number of electrodes and the sampling rate of ADC.

Various wireless data communication methods have been proposed in the literature. These methods can mainly be classified into two groups. Data communication

© Springer Nature Switzerland AG 2020
K. Türe et al., *Wireless Power Transfer and Data Communication for Intracranial Neural Recording Applications*, Analog Circuits and Signal Processing,
https://doi.org/10.1007/978-3-030-40826-8_4

on the power line by changing wireless power transfer parameters is one solution for the remotely powered systems [2, 3]. The other way to transmit data is by implementing a dedicated transceiver on the implant and an external station [4, 5]. The appropriate solution for the wireless data transmission can be decided by considering the power budget, required data rate, and additional antenna space for the dedicated transceiver. For low power budget, low data rate, and limited implant size applications, communication data from an external station to an implanted system can be transferred by the magnetically coupled coils, which are commonly used to transmit power and data. However, the maximum data rate achievable using the powering link is limited and in some applications such as neural monitoring, utilization of dedicated transmitter becomes inevitable.

The development of microelectrode arrays addresses high spatial and temporal resolution requirements. They perform a simultaneous and high-density monitoring of action potentials in distributed brain areas [6]. Data acquired from hundreds of electrodes can quickly reach up to hundreds of megabits per second (Mbps). Those constraints obligate wireless transmitters with high data rate at low power dissipation for neural monitoring applications. However, the configuration of the implanted device does not demand such a high data rate for downlink communication. Accordingly, different solutions are presented for each of the data communication links.

This chapter begins by presenting a downlink communication solution based on modulating the remote powering signal. It will then go on to methods for uplink data transmission. For uplink communication, different circuit topologies for narrowband and ultra-wideband transmitters are presented. Finally, the conclusion gives a summary of the findings.

4.2   Downlink Data Communication

Data transfer from the external station to the implant can be directly performed by modulating the powering signal source. A simple modulation scheme reduces complexity and power dissipation in both the transmitter and receiver sides. The most commonly used amplitude shift keying (ASK) modulates the amplitude of the powering signal for transmitting "0" and "1" bits [7, 8]. The application of frequency shift keying (FSK) and phase shift keying (PSK) is to send data to inductively powered implant by changing the frequency [9, 10] and phase [11, 12] of the inductive link, respectively.

The magnetic coupling between the powering and the implanted coils changes because of misalignments. Since the remote powering performance and also the downlink communication reliability depend on the amplitude of the induced voltage at the implanted side, they are also affected by the change of the magnetic coupling. This book presents a low power pulse position modulation (PPM) modulator and demodulator that guarantee reliable communication with an efficient remote powering.

The changes in received signal occur when there is a misalignment between the external and implanted powering coils. However, for reliable communication between the external base station and the implanted system, these voltage changes should not have an impact on the data transmission. ASK modulation scheme is widely used for data communication in remotely powered applications due to its simplicity regarding design and low power consumption compared to the frequency and phase modulations. The data is encoded by specific modification of the amplitude of the remote powering signal. Most of the ASK demodulators use an envelope detector followed by a comparator in order to detect changes in the magnitude of the powering signal [13, 14]. The envelope detector takes the powering signal and provides the amplitude changes in the signal as an output. The comparator receives the output of the envelope detector and decides whether the data is "0" or "1" by comparing it with a reference voltage.

The ASK modulation index (mi) can be defined as

$$mi = \frac{V_{\max} - V_{\min}}{V_{\max} + V_{\min}} \tag{4.1}$$

where $V_{\max}$ is the maximum amplitude of the signal and $V_{\min}$ is the minimum amplitude, which corresponds to data bit "1" and "0," respectively. The mi is a significant parameter that affects bit error rate (BER) and the reliability of the communication. The increasing mi indicates that the voltage difference in the amplitude of the powering signal for different data values is increasing. It makes detection easier and decreases BER. However, high mi implies high amplitude change in the powering signal, and it reduces the powering performance. This causes a trade-off between communication and powering performances. Besides, the difference in the powering signal due to the lateral misalignments, as shown in Fig. 4.1, makes the communication unreliable. The demodulator cannot distinguish the variations in the powering signal, which are because of the motion of the external unit or the modulation of the communication signal. One way to solve this issue is to select very high mi, close to 1, that creates more changes in the powering signal. However, as it is mentioned, this reduces the powering performance. Therefore, an efficient encoding scheme that maximizes the powering performance and minimizes BER at the same time needs to be designed.

Another technique for defining the communication data bits is the use of pulsewidth modulation (PWM). The different bits in the communication with PWM are represented by different duty cycles of constant amplitude pulses. Figure 4.2 shows the mentioned two modulation types and corresponding data bit values. The time variables t_{cycle}, t_H, and t_L are the period of one-bit data, the duration of the high voltage, and the duration of the low voltage, respectively. As seen from the figure, the data in the PWM is determined by the t_H and t_L. In case of t_H is longer than t_L, the data represents "1," or otherwise "0."

The effect of the mi value on the performance of the powering operation and the reliability of the communication is discussed before. A new scheme of modulation is required to find a good compromise between powering and communication. As

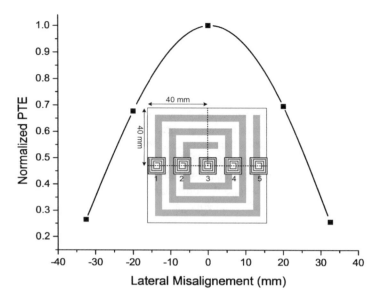

Fig. 4.1 Normalized power transfer efficiency versus lateral misalignment

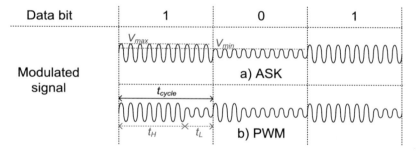

Fig. 4.2 Two encoding schemes and modulated communication signal

a solution to this problem, the pulse position modulation (PPM) encoding scheme was proposed by Kilinc et al. [15]. The aim of this PPM scheme is to minimize the duration of low voltage amplitude state in order to keep the powering stable while making the communication reliable with changing the timing of the low voltage amplitude. Figure 4.3 shows the ASK, PWM, and PPM encoding schemes when mi is 1 and their effects on the regulated voltage at the implanted system. Compared to the ASK and PWM, in the PPM scheme, the data bits "0" and "1" are encoded by the position of the interruption of the powering signal. In this way, communication becomes independent of the t_L duration.

Beyond the time t_L, several other parameters affect the remote powering stability such as the consumed power from the regulated voltage, the rectified

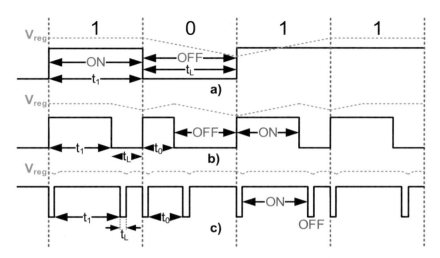

Fig. 4.3 ASK (**a**), PWM (**b**), and PPM (**c**) encoding schemes and effects on the remote powering operation

voltage level, and used capacitor values at the output of the rectifier and voltage regulator. Increasing current consumption from the regulated voltage decreases the discharging time of the storage capacitor. In general, there is also a storage capacitor at the output of the rectified voltage and its voltage across the terminals depends on the remote powering signal. High swing of the induced voltage creates a high voltage at the output of the rectifier and makes the discharging time of the capacitor longer. Another approach for keeping the voltage level at a sufficient value during a longer duration is to use a higher storage capacitor. Figure 4.3 also shows the affected regulated signal in blue due to the powering signal cuts. These effects are illustrated with the consideration of the same powering voltage level, the equal current drawing from the regulated signal, and the identical storage capacitors. Smaller t_L makes the powering signal more stable, and the PPM encoding shows an excellent solution to remote powering and communication trade-off compared to the other two methods.

4.2.1 PPM Modulator

PPM modulator takes the data at every clock cycle and determines the corresponding position of the power interruption. Figure 4.4 shows the block diagram of the external base station for powering and downlink communication. The external coils are driven by a class-E power amplifier (PA) at powering frequency generated by the oscillator. The mi value is selected as 1 in order to increase the accuracy of the communication. The changes in the communication signal are implemented by

Fig. 4.4 Block diagram for
external base station

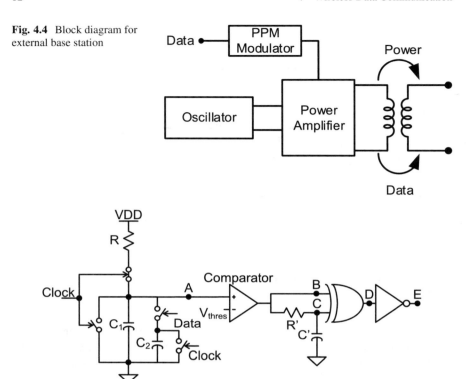

Fig. 4.5 Schematic of PPM modulator

switching on and off the PA supply voltage by the output of the PPM modulator
unit. Figure 4.5 shows the more detailed circuitry for the PPM modulator.

The data input value regulates the capacitance at the RC network and so the
charging time. The data controls the total capacitance in the RC network. The time
t_0 determined by the RC_1 network, while t_1 is set by the $R(C_1 + C_2)$ time constant.
The clock signal discharges the capacitors at every beginning of a cycle. When the
accumulated voltage at node A reaches the threshold voltage of the comparator, it
generates a transition from low to high as an output. By using an $R'C'$ delay line
and an XOR gate, a pulse is created. The $R'C'$ delay controls the created pulse
width. This pulse generation occurs at every output changes of the comparator.
Therefore, there will be no need for additional circuitry for creating another pulse
at the beginning of every cycle. As a final stage, an inverter is used for converting
the positive pulse generated by the XOR gate to the supply voltage of the PA. By
cutting the supply voltage of the PA, the current passing through the external coil is
interrupted for the duration of the pulse and so the induced voltage at the implanted
system. Figure 4.6 shows the operation of the PPM modulator, for data bits "0"
and "1."

Fig. 4.6 Operation of PPM
modulator

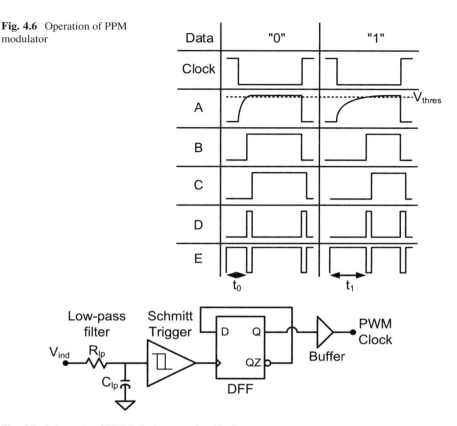

Fig. 4.7   Schematic of PWM clock generation block

4.2.2   PPM Demodulator

The PPM demodulator is composed of three blocks: PWM clock generation, counter block, and data determination block. The PWM clock generation block [7] is shown in Fig. 4.7. The PPM encoded induced voltage, V_{ind}, is filtered by an RC low pass filter. The high-frequency component of V_{ind} is filtered out and only changes because of encoding remains. Since the V_{ind} voltage can vary due to not only the encoding but also the movement of the external base station and other environmental fluctuations, using a Schmitt trigger with hysteresis prevents incorrect state changes and malfunctioning. Therefore, a Schmitt trigger with 150 mV hysteresis is used for the elimination of ripples at the input and provides a rail-to-rail signal as an output. D flip-flop (DFF), following the Schmitt trigger, creates a PWM clock that changes its state at every rising edge of the Schmitt trigger output. As a consequence of this block, the PPM decoded V_{ind} is converted into a PWM clock.

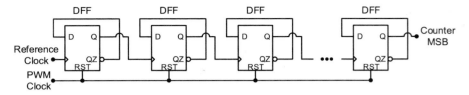

Fig. 4.8 Schematic of counter block

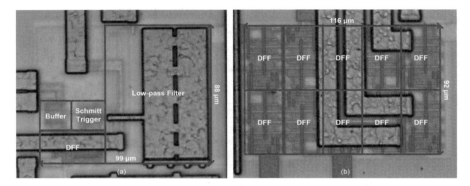

Fig. 4.9 Micrograph of (**a**) PWM clock generation and (**b**) counter

The counter block is for measuring the high voltage times, t_1 and t_0. The schematic diagram of the counter block [15] is shown in Fig. 4.8. Series connected DFFs imply the counting action. Each DFF divides the input frequency by half. All of the DFFs are enabled by an active low signal generated by the PWM clock so that the counter keeps its operation for a time interval t_1 and t_0. The half period of the counter, $t_{counter}$, and the other two time parameters t_1 and t_0 are selected such that they satisfy the inequality $t_1 > t_{counter} > t_0$. Accordingly, the output of the last DFF, most significant bit (MSB) of the counter, changes its state from "1" to "0" if the duration is t_1. Contrarily, if the duration is t_0, the MSB does not change its state and keeps at "1" during the time $t_{counter}$. Therefore, the data bits "0" and "1" can be decoded by monitoring the changes in the output state of the last DFF. A ring oscillator is used for the generation of the reference clock for the counter block. The circuits are fabricated in a $0.18\,\mu$m CMOS process. Figure 4.9 shows the micrograph of the PWM clock generation and the counter blocks [15].

As a last part of the demodulator, the data determination block monitors the changes at the MSB of the counter and compares the change time with the PWM clock signal. The schematic diagram for the data determination block is shown in Fig. 4.10. The falling edge detector, built by three inverters and a NOR gate, detects the changes from high to low voltage in the MSB of the counter and creates a short positive pulse. This short pulse is compared with the PWM clock signal with the help of a NAND gate. If the data is "1," the NAND gate gives high voltage as an output for the duration of the pulse. Otherwise, it stays at zero voltage for

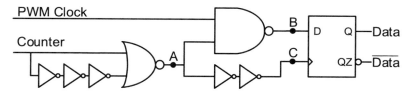

Fig. 4.10 Schematic diagram for data determination block

Fig. 4.11 Operation of data determination block

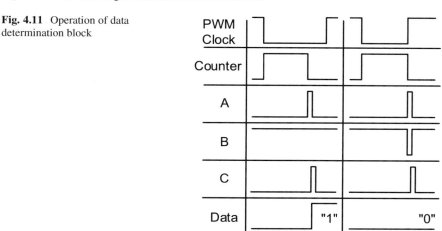

the same time interval. This decision is stored in a DFF, which is activated by the delayed output of the falling edge detector. The operation of the data determination block is depicted in Fig. 4.11 for both data bit "0" and "1" cases. For the data bit representation in the last row, it is assumed that the previous bit is "0."

4.2.3 Measurement Results and Discussion

The power transfer efficiency is strongly dependent on the quality factors of the coupled coils [16]. Therefore, for an efficient power transfer, the high-quality factor is preferred for the coils. On the other hand, the coils with high-quality factors have a low response time that limits the communication bandwidth. Accordingly, the interruption in the powering signal for the data communication needs to be selected carefully such that it will be long enough to create a drop in the induced voltage while not disturbing the remote powering performance. More precisely, the voltage filtered by an RC low pass filter needs to be lowered enough to activate the Schmitt trigger. This condition is also a design criterion for the lower point threshold of the Schmitt trigger. The higher point threshold of the Schmitt trigger needs to be determined so that it will not require a signal for triggering, which is higher than the minimum powering signal for the optimum operation.

The storage capacitor should fulfill the energy loss during the cut in the powering signal. The required capacitance value is determined by the power consumption from the regulated voltage and duration of the power cut. The usage of a big capacitor is favorable regarding the powering of the system. However, the required time to charge the capacitor increases with the capacitance and restricts the high data rate communication. After the analysis of all the limitations of the stable powering and reliable communication, one-bit duration (t_{cycle}) is chosen as 120 μs that corresponds to 8.33 kbps data rate. The duration for the bit "1" and the bit "0" is selected as 100 and 30 μs, respectively. The ring oscillator provides a reference clock signal at 1.45 MHz frequency. The eight stages of DFFs use this clock signal in the counter block to measure 88.3 μs of high-level duration. The communication data is decoded as a result of the comparison with the 88.3 μs reference duration.

Figure 4.12 shows the measured waveform of PPM communication for the bit "1" (top) and the bit "0" (bottom). The modulator output voltage for the supply of the class-E PA that drives the external coil is shown in red. 7 μs duration of a supply

Fig. 4.12 Measured waveform of PPM communication for (**a**) the bit "1" and (**b**) the bit "0"

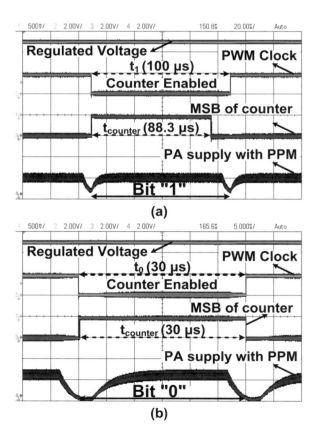

cut is chosen in order to create enough power cut that activates Schmitt trigger. The 1.8 V regulated voltage on the implanted side (as shown in yellow) stays constant during the data communication as it is desired. The output of the clock generation block generates the PWM clock by detecting the power cut, as shown in green. The PWM clock is used to enable or disable the DFFs, which are turned on by the active low signal. "Counter Enabled" duration is measured by the counter block. If this duration is longer than $88.3\,\mu s$, the output of the counter block (as shown in blue) changes from logic "1" to logic "0"; otherwise, the counter keeps its logic state at "1." Accordingly, t_1 and t_0 durations are decoded as the bit "1" (on the top) and the bit "0" (on the bottom), respectively.

Figure 4.13 shows the measurement results that use the same circuitry except for the ASK modulation scheme with $mi = 1$ for the communication. The regulated voltage in ASK modulation has fluctuations of about 100 mV because of the remote power cuts, while it is not the case for the PPM scheme. Moreover, ASK modulated communication requires a higher supply voltage for the PA to compensate for the loss of sending bit "0." As it is discussed through the paper, the PPM scheme shows better performance for the remote powering. The proposed PPM scheme shows a promising and reliable solution for sending downlink data to an implanted system without being affected by the misalignments in time.

Fig. 4.13 Downlink modulation scheme effect on the remote powering

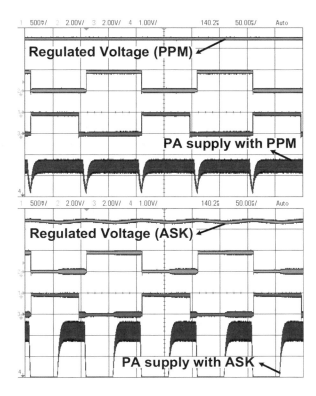

4.3   Uplink Data Communication

Following the realization of the downlink data communication to program the controller unit in the implanted system, there is a need for transferring the measurement results to the external base station. Thanks to the improvements in electrodes, sensors, and microelectronics, the need for quality and quantity of extracted information has been on the rise. It is thus expected that recording systems improve performances, especially in the spatial and temporal resolutions. The development of microelectrode arrays addresses high spatial and temporal resolution requirements. Data acquired from hundreds of electrodes can quickly reach up to hundreds of megabits per second (Mbps). Those constraints dictate two possible scenarios. One solution is the compression of the information extracted from neural signals by merging new mathematical theory and computational methods in hardware. Compressive sensing reduces the amount of transmitted data, while still allowing robust, but complex, off-line reconstruction of the original signal. The other way is to designing wireless transmitters with a high data rate at low power dissipation for neural monitoring applications without a need for compression.

In recent years, several implanted neural monitoring systems have been developed to enable wireless neural activity recording. For example, in the 64-channel wireless micron-scale electrocorticography (μECoG) solution presented in [17], Miller-encoded serialized data is transmitted at 1 Mbps thanks to a backscattering modulator. As another example, a binary-phase-shift-keying (BPSK) modulated transmitter at 2.4 GHz with the capability of delivering data up to 8 Mbps from 8 channels was developed [18]. Further, a neuroelectrical monitoring CMOS system utilized a Manchester-encoded FSK modulation scheme with a carrier frequency at 916.4 MHz and a data rate of 1.5 Mbps across a distance of 10 m [19].

Higher data rates compared to the conventional narrowband transmitters can be achieved by using pulse-based communication methods instead of carrier-based radios with the cost of limited transmission range. In this kind of wide-band data transmission, each symbol is encoded by a train of short pulses. Kiani and Ghovanloo [20] have developed a near-field data transmission across inductive telemetry links based on pulse harmonic modulation with a data rate of 20 Mbps. Impulse radio ultra-wideband (IR-UWB) technology has drawn attention for low power, high data rate communication for implanted systems [21–23] even though high absorption loss of body tissue limits the transmission range.

In the presented book, two different kinds of wireless data transmitters are designed for mentioned two possible solutions. The first narrowband transmitter is operating together with the compressive sensing block. The narrowband transmitter has limited capability regarding the data rate but provides an extended range of operation. On the other hand, the other type of transmitter, namely the ultra-wideband transmitter, is able to send an excessive amount of data while suffering from the communication distance. Depending on the need during the presurgical analysis, one of the transmitters can be activated by programming the embedded controller unit with the help of the downlink data transmitter, as mentioned earlier.

4.3.1 Narrowband Transmitter

Data generated by the sensor network is transferred wirelessly to an external base station using different types of modulation techniques. During the data transmission, "0" and "1" bits can be coded by changing the signal parameters such as its amplitude, frequency, and phase. Every modulation technique has its advantages and disadvantages. In an implantable system, in addition to the size and weight restrictions, there is also a limitation for the power budget. Accordingly, on-off keying (OOK) modulation is a promising solution for low power consumption because it does not consume power for "0" bits, unlike the other modulation methods.

Moreover, OOK modulation is very simple to implement concerning the design perspective since it requires only the activation and deactivation of the oscillator. However, although OOK modulation consumes less power and has a simple structure, its data rate is reduced due to the slow nature of transitions from ON state to OFF state and vice versa. In this work, the methods for the optimization of OOK modulation regarding data rate and their results are presented.

4.3.1.1 Transmitter Architecture

In order to analyze the rising and falling time of an oscillator, a cross-coupled symmetrical voltage controlled oscillator (VCO) whose schematic is presented in Fig. 4.14 is chosen as a reference oscillator. This type of oscillator is selected since it provides balanced output and shows a better phase noise performance at a given power dissipation compared to the asymmetrical version of it [24]. The resonance frequency of the oscillator is determined by the tuning capacitors and the inductance of a loop antenna, which is implemented as an off-chip. The antenna is directly connected to the differential pins of the cross-coupled VCO. The parameters for the tuning capacitors and the loop antenna are selected, such that it oscillates in the MedRadio band at the 416 MHz frequency.

The OOK encoded signals at the output of the oscillator is generated by turning ON/OFF the current source of the cross-coupled VCO. The modulated bias current is multiplied by a current mirror, and this multiplication introduces some delay between the data and oscillator output. When the current source is switched ON, the output of the VCO starts to build up exponentially. This similar exponential behavior of the VCO is obtained during the turn-off procedure of the oscillator. Figure 4.15 shows the simulation results for the response of the oscillator designed in $0.18\,\mu m$ technology to the applied data signal.

Fig. 4.14 Schematic of the
reference cross-coupled
voltage controlled oscillator

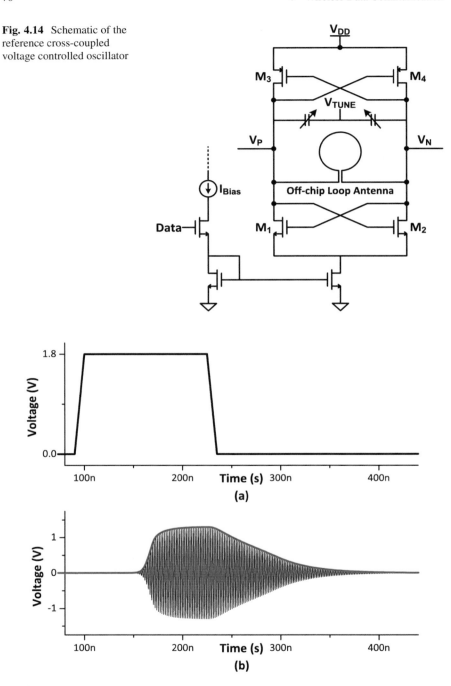

Fig. 4.15 (**a**) Applied data signal to the oscillator. (**b**) Simulated output response of the oscillator

Fig. 4.16 Equivalent circuit
of the cross-coupled VCO

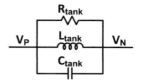

Figure 4.15a represents the message signal and Fig. 4.15b shows the output
voltage response and its envelope to the data for modulated 20 μA bias current. The
exponential response of the oscillator can be analyzed by considering the equivalent
circuit of the VCO as shown in Fig. 4.16 and the corresponding equivalent parameters as

$$L_{tank} = L_{loop} \tag{4.2}$$

$$R_{tank} = 1/(g_{tank} + g_t + g_L) \tag{4.3}$$

$$C_{tank} = C_{NMOS} + C_{PMOS} + C_L + C_{load} + C_V \tag{4.4}$$

where g_{tank} is the sum of the output conductance of NMOS and PMOS, g_t is the
parasitic conductance of the tuning capacitors, g_L is the parasitic conductance of
loop antenna, C_{NMOS} and C_{PMOS} are the parasitic capacitance of cross-coupled
transistors, C_L is the parasitic capacitance of the loop antenna, and C_{load} is the
parasitic capacitance at the load of the oscillator [25]. The waveform of the oscillator
output, $V_{out} = V_P(t) - V_N(t)$, for a unity bias current can be expressed by [26]

$$V_{out} = \frac{e^{-\xi \omega_0 t}}{\sqrt{1 - \xi^2}} \cos \left[\omega_0 \sqrt{1 - \xi^2} t - \tan^{-1} \left(\frac{\xi}{\sqrt{1 - \xi^2}} \right) \right] \tag{4.5}$$

where

$$\xi = \frac{1/(2 R_{tank} C_{tank})}{1/\sqrt{L_{tank} C_{tank}}}. \tag{4.6}$$

Equations 4.5 and 4.6 express that the time constant of oscillation envelope
is $2 R_{tank} C_{tank}$ while ω_0 is stated as $1/\sqrt{L_{tank} C_{tank}}$. This relation shows that
the shortest rising time/falling time of the oscillator can be achieved by the
smallest R_{tank}. Since R_{tank} includes the output conductance of the cross-coupled
transistors, increasing bias current increases the output conductances and decreases
the equivalent resistance of the tank, R_{tank}. However, increasing bias current also
increases the power consumption of the VCO. Therefore, in this work, a higher bias
current is applied during the rising time of the oscillator in order to obtain faster
switching with keeping the additional power consumption minimum. For the falling
time of the oscillator, the equivalent resistance of the tank is minimized by adding a

very small resistance in parallel to R_{tank}. The details of these methods for increasing the switching time of the oscillator are presented in the next subsection.

4.3.1.2 Optimization of Rising Time

Figure 4.17 shows the OOK modulated data signal for a conventional solution and possible methods used to speed up the rising of the oscillator output. Figure 4.17a depicts the conventional OOK modulated signal. It remains high from time t_1 to t_2, an interval that must be larger than the delay time featured in Fig. 4.15b. The other two figures show waveforms with boosted currents at the turn-on time. Figure 4.17b foresees an extra constant current. Figure 4.17c uses a pulse fading exponentially added to the bias current. Since the current required to sustain oscillation has a value lower than the one necessary to start oscillation, the current amplitude after the pulses can be brought close to the minimum. The boosted current in Fig. 4.17b requires another constant current source for the biasing and monostable circuit that controls the timing of the start-up current. As an alternative to the constant biasing current, an exponentially decaying bias current is proposed since it requires fewer components to generate current and to control the time constant of the decay. Figure 4.18 shows the required circuit for the generation of the current in Fig. 4.17c.

A differentiator circuit generates the decaying bias current. The network composed of R_{bias} and C_{bias} creates the derivative of the inverted data signal at the gate

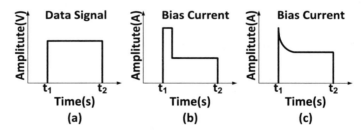

Fig. 4.17 (**a**) Applied data signal to the oscillator. Modulated bias current with additional (**b**) constant current and (**c**) exponentially decaying current

Fig. 4.18 Schematic for generating exponentially decaying current

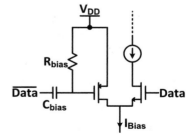

of the PMOS transistor, and the transistor produces a current proportional to the derivative. Accordingly, the generated current rises with the rising of the data signal and reaches a peak when there is no more change in the data signal. The peak of the generated current flow by the PMOS transistor strongly depends on the slope of the data signal when it is rising. After the data signal reaches its stable form, the current starts to decay with a time constant of $R_{bias}C_{bias}$. Since the peak current depends on the rise time of the pulse controlling the gate of the p-channel transistor and the decay depends on the $R_{bias}C_{bias}$ time constant, suitable control of variables leads to an optimal rise time of the oscillator while keeping limited the extra power consumption.

4.3.1.3   Optimization of Falling Time

As evident from Fig. 4.15b the fall time also limits the speed of data transmission. The solution used here foresees the simple use of a switch controlled by the data. Thanks to the low on-resistance of the switch, the time constant of the tank $R_{tank}C_{tank}$ becomes very small, thus stopping the oscillation in a very short time. Therefore, a switch with low ON resistance is connected to the differential output pins of the oscillator, as shown in Fig. 4.19. The added switch introduces a small capacitance at the output of the oscillator, which causes a shift in the oscillation frequency. Since the frequency change is minimal, the variation can be compensated by the tuning voltage of the varactors. The dimensions of this transistor have been chosen so that it stays in normal operation. The optimized oscillator concerning rising and falling times of the output and its simulation results are presented in the next section.

4.3.1.4   Post-Layout Simulation Results

The optimizations for rising and falling times of the oscillator output are applied to the reference circuitry and represented in Fig. 4.20. The optimized VCO is designed in 0.18 μm technology. The constant current bias is kept the same in order to obtain the same steady-state amplitude with the reference circuitry. The time constant $R_{tank}C_{tank}$ for the decaying bias current is set to 4.5 ns for optimization of data rate and power consumption. The post-layout simulated waveform of the reference and optimized VCO is shown in Fig. 4.21.

Fig. 4.19 Schematic of the switch added to the differential output pins of the oscillator

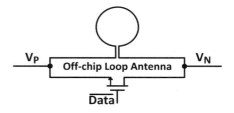

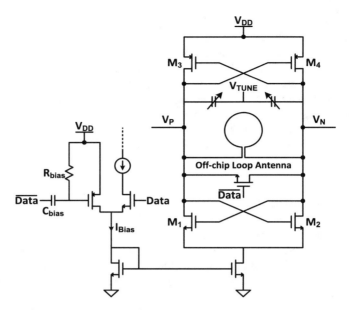

Fig. 4.20 Schematic of the optimized cross-coupled voltage controlled oscillator

The proposed optimization improves the rising time and the falling time of the oscillator significantly. The maximum data rate achieved by the oscillator increased drastically. Figure 4.22 shows the optimized performance of the oscillator with a 25 Mbps data rate. These optimization methods provide a significant increase in the timing properties of the oscillator by introducing a few additional components.

4.3.1.5   Measurement Results

The proposed narrowband transmitter was designed in UMC 180 nm process technology. The transmitter occupies an area of $510\,\mu m \times 160\,\mu m$. The micrograph of the fabricated transmitter is shown in Fig. 4.23.

The external loop antenna is designed to maximize the radiation efficiency at the transmission frequency. The radiation efficiency of a loop antenna is expressed as [27]

$$E_{loop} = \frac{R_{rad}}{R_{rad} + R_{ohmic}} \tag{4.7}$$

where R_{rad} is the radiation resistance and R_{ohmic} is the resistance of the loop antenna. It is expressed that the radiation resistance is proportional to the area of the loop and resistance of the loop decreases with the increase of the metal wire width [28].

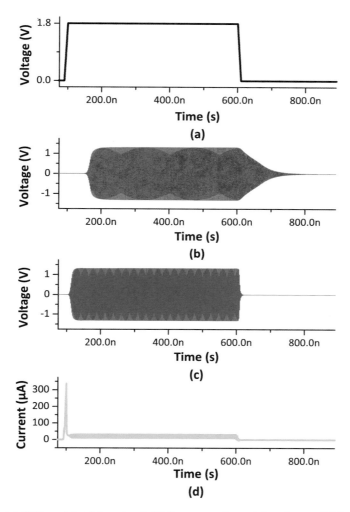

Fig. 4.21 (**a**) OOK modulated data signal. (**b**) Output waveform of the reference VCO. (**c**) Output waveform of the optimized VCO. (**d**) Generated bias current for the VCO

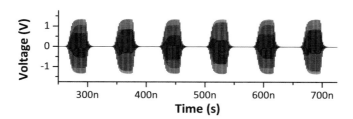

Fig. 4.22 Output waveform of the oscillator with 25 Mbps OOK signal

Fig. 4.23 Micrograph of the narrowband transmitter

Fig. 4.24 Fabricated loop antenna to be used in narrowband transmitter

The dimensions of the Burr hole restrict the implantable system. The diameter of the antenna is chosen as 15 mm to fit inside the hole and maximization of the radiation resistance. It has been noted that increasing the width of the copper wire after a particular value does not affect the resistive losses since the skin effect dominates it. Therefore, the width of the copper wire, which constitutes the loop antenna is selected as 1 mm. The loop antenna is fabricated on the FR4 substrate with a thickness of 0.3 mm using 18 μm thick copper. The inductance of the loop antenna is measured as 35 nH. Figure 4.24 shows the fabricated loop antenna for the narrowband transmitter.

In order to analyze the performance effects of the additional circuits in traditional VCO, the circuits without improvements and with improvements are measured in the same conditions. A pulse of 500 ns duration is applied to both of them as the data signal. The supply is chosen as 1.8 V for maximum oscillation amplitude. Their measured output waveforms are depicted in Fig. 4.25. Waveforms show that the improvements in propagation times are significant. Table 4.1 summarizes the measurement results in detail.

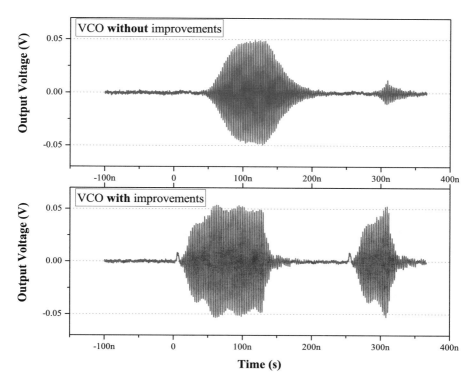

Fig. 4.25 Output waveform of VCO without improvements (top) and with improvements (bottom) for applied "1" bit data between 0 and 125 ns and from 250 to 310 ns

Table 4.1 Design parameters of inductive link coils

	VCO without improvements	VCO with improvements
Rise time (ns)	38	39
Fall time (ns)	52	15
$0 \rightarrow 1$ Propagation time (ns)	90	52
$1 \rightarrow 0$ Propagation time (ns)	64	30
Maximum data rate (Mbps)	6.5	12.2
DC power consumption (μW)	509	544
FoM (pJ/bit)	78	44

Rise time and fall time correspond to the transition of the oscillation envelope, respectively, from 10 to 90% and from 90 to 10% of its maximum steady-state amplitude. Propagation time indicates the time requiring to settle down of the oscillator when there is a change in the data signal. Achievable maximum data rate and DC power consumption are obtained by considering propagation times with the assumption of the same durations and probability for "0" and "1" bits. The figure of merit (FoM) is defined by the consumed energy for transmitting one bit of information at the maximum achievable data rate. The proposed optimization improves the maximum data rate achieved by the oscillator by 87% while the DC power consumption increases 6.7%, which yields an improvement of 1.8 times in figure-of-merit.

4.3.2   Ultra-Wideband Transmitter

Impulse radio ultra-wideband (IR-UWB) is a promising technique based on the transmission of short pulses, and it is very efficient for low range applications that require a high data rate. In 2002, the Federal Communications Commission (FCC) approved and limited the maximum effective isotropic radiated power (EIRP) to -41.3 dBm/MHz for bandwidth between 3.1 and 10.6 GHz [29]. There are three main methods for generating UWB pulses: combining filtered edges [21], combinations of pulses [22], and modulating a local oscillator [23]. The filtered combined edges method is desirable for its low power consumption. However, generating FCC-compliant pulses requires an area of expensive filter implementation [21]. Another low power IR-UWB method combines short pulses using highly digital circuits. However, due to the significant effect of variations between stages on pulse duration and its power spectral density (PSD), this method requires complicated calibrations [30].

A ring oscillator can be used as the local oscillator (LO) to be modulated for pulse generation, but it suffers from instability in frequency. However, since the communication is pulsed based and the frequency spectrum is broad, frequency instability does not prevent ring oscillators from being used in IR-UWB transmitters. The drawback of LC oscillators as an alternative to ring oscillators is the higher silicon area consumption and power requirements [31]. Nevertheless, the mentioned limitations of the spiral inductor in a conventional on-chip LC tank are overcome by replacing it with an active inductor topology. This work presents two different high data rate, energy and area efficient LO-based IR-UWB transmitters; one of them uses a ring oscillator for the generation of the carrier frequency. The second IR-UWB transmitter benefits of the active inductor topology and eliminates the massive area occupation of the spiral inductor in LC oscillator.

It has been shown in [32] that a Gaussian pulse shape optimizes for maximum bandwidth and minimum sidelobe leakage to fulfill FCC limits. However, Gaussian pulses require complex circuit design and high power consumption. Triangular shaped pulses can be seen as a favorable alternative, considering the design simplicity, power, and area requirements. In this work, the shapes of the pulses are selected as triangular to gain in design with minimal lost in the PSD spectrum.

4.3.2.1   Ultra-Wideband Transmitter Based on Ring Oscillator

Figure 4.26 shows the schematic block diagram of the IR-UWB transmitter. The small number of circuit elements makes the design simpler and area occupation minimal. The core of the transmitter is based on the current-starved ring oscillator,

Fig. 4.26 Schematic of the IR-UWB transmitter based on ring oscillator

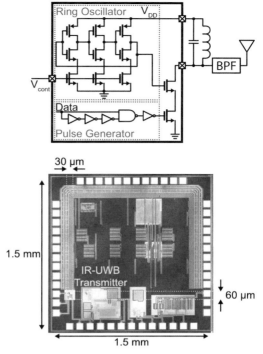

Fig. 4.27 Micrograph of the IR-UWB transmitter based on ring oscillator

which generates output in the range of 3.5–4.5 GHz frequency. The control voltage provides flexibility in selecting the oscillation frequency of the ring oscillator by adjusting the bias current. The pulse generator block creates short pulses at the rising edges of the data signal. The output of the ring oscillator and the pulse generator is mixed with cascode connected transistors. The drain of the transistor driven by the ring oscillator is connected to the external resonator circuit formed by an inductor and a capacitor. Before the 50 Ω UWB antenna, a band-pass filter (BPF) centered at 4 GHz is used in order to satisfy the FCC regulation. For the transmission of the generated IR-UWB pulses, miniaturized, flexible, and polarization-diverse UWB antenna presented in [33] can be adopted.

The proposed IR-UWB transmitter is fabricated in UMC 180 nm process technology and it occupies 60 μm × 30 μm area. The micrograph of the fabricated transmitter is shown in Fig. 4.27. Figure 4.28 shows the measured output waveform of the implemented IR-UWB transmitter with 250 MHz pulse repetition rate. The maximum peak-to-peak amplitude of the measured pulse is 111 mV, while its duration is 2.2 ns. Figure 4.29 depicts the measured power spectral density of the transmitter and FCC regulation. The triangular envelope of the output waveform suppresses the side-lobes and the measured spectrum fully meets the FCC mask. When the pulse repetition frequency is 250 megapulses per second (Mpps), the complete IR-UWB transmitter consumes 11.3 mW power, which corresponds to 45.2 pJ/pulse. The high throughput of the IR-UWB transmitter makes it possible to buffer the raw data and transmit it in several bursts.

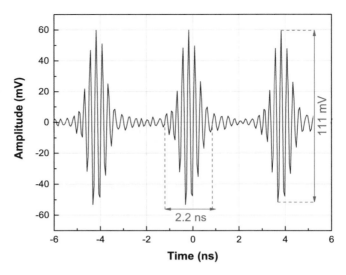

Fig. 4.28 Transient pulses of the IR-UWB transmitter at 250 Mpps

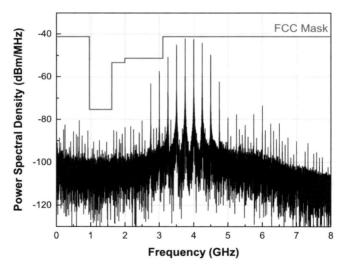

Fig. 4.29 Power spectral density of the IR-UWB transmitter

4.3.2.2   Ultra-Wideband Transmitter Based on Active Inductor

To decrease the large area consumption associated with conventional spiral CMOS integrated inductors while retaining the start-up behavior of a regular LC tank oscillator, an active inductor can be used. An active inductor is built solely out of transistors and capacitors, thus removing the need for a large area integrated inductor. A significant drawback of the usage of active inductors in RF systems is

Fig. 4.30 Schematic of the proposed pulsed active inductor oscillator

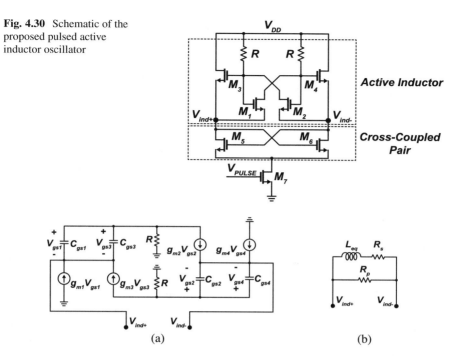

Fig. 4.31 (**a**) Small-signal model of the active inductor. (**b**) Simplified equivalent model of the active inductor

that they have much worse phase noise performance than their large area inductor counterparts for the same power consumption, as shown in [34]. However, since the IR-UWB technique is pulse based, phase noise is of no concern and active inductors become very attractive.

The schematic of the LO based on an active inductor is shown in Fig. 4.30. The active inductor topology introduced by Lu et al. [35] is realized by the transistors M_1–M_4 and two resistors R. Drain-source and drain-gate parasitic capacitances of the cross-coupled pair of transistors, M_5 and M_6, constitute an LC tank together with the active inductor. Moreover, M_5 and M_6 provide a sufficient negative conductance to satisfy the Barkhausen criterion by compensating the loss of the LC tank and ensure start-up and sustained oscillation. By introducing a current source at the bottom of the LO topology, the oscillator can be turned ON or OFF. Since an active inductor is used in cooperation with the parasitic capacitances of the transistors, an LC start-up and slow-down behavior can be realized without the need for large area integrated passive inductors.

The selection of transistor dimensions, the resistance value, and determination of the oscillation criterion requires a small-signal analysis. The small-signal model of the active inductor and its equivalent model are shown in Fig. 4.31a and b, respectively. Symmetry in the circuit yields the input impedance as

$$Z_{in} = \frac{2\left[jw(C_{gs1} + C_{gs3})R - g_{m1}R + 1\right]}{g_{m1} + g_{m3} + jw(C_{gs1} + C_{gs3})}. \qquad (4.8)$$

For two criteria, $2g_{m1} + g_{m3} > R^{-1}$ and $g_{m1} < R^{-1}$, the parameters in the equivalent model of the active inductor are given by

$$R_p = 2R \qquad (4.9)$$

$$R_s = \frac{2R(1 - g_{m1}R)}{(2g_{m1} + g_{m3})R - 1} \qquad (4.10)$$

$$L_{eq} = \frac{2(C_{gs1} + C_{gs3})R^2}{(2g_{m1} + g_{m3})R - 1}. \qquad (4.11)$$

This equivalent model indicates the requirement of the negative conductance provided by the cross-coupled pair of M_5 and M_6. The transconductance g_{m5} needs to be higher than R_p^{-1} to guarantee the start-up of the oscillation. Transistor M_7 is sized such that it provides the necessary bias current to the LO for the desired oscillation and start-up behavior for incoming short pulses to its gate.

Figure 4.32 shows the proposed IR-UWB transmitter which includes a pulse generator, an active inductor-based oscillator, and a single-stage amplifier constituted by transistors M_8 and M_9 of size 50/0.18 μm. The pulse generator is composed of an inverter chain and an AND-gate logic. In the inverter chain, current-starved inverters are used to provide the flexibility of controlling transition times and delays by an external voltage, V_{ctrl}. This feature provides a control over the width of the generated pulse. The AND-gate logic circuit needs to be sized such that the output stage can drive the gate of transistor M_7 for a desired bias current of the LO. The

Fig. 4.32 Proposed IR-UWB transmitter based on active inductor

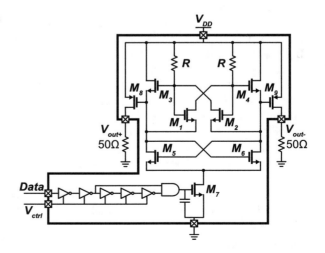

Table 4.2 Circuit parameters
of the local oscillator

	W/L (μm/μm)
M_1, M_2	25/0.18
M_3, M_4	25/0.18
M_5, M_6	80/0.18
M_7	150/0.18

Fig. 4.33 Microphotograph
of the IR-UWB transmitter
based on active inductor

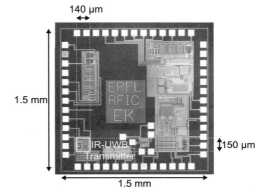

equivalent inductance value L_{eq} is determined as 1.8 nH for the desired operation
of the IR-UWB transmitter. The value of the resistor R is chosen as 125 Ω and the
transistor sizes used for the active inductor-based oscillator design are reported in
Table 4.2.

The differential output of the oscillator needs to be isolated from the loads to
prevent the highly sensitive active inductor from being affected by the load. For
isolation of the oscillator and amplification of the output waveform, two single-
stage amplifiers were utilized, and their drains directly drive the differential loads,
as shown in Fig. 4.32.

The proposed IR-UWB transmitter is fabricated using UMC 180 nm MM/RF
process technology. The total active area occupies only 150 μm × 140 μm. The
fabricated chip is shown in Fig. 4.33, where the IR-UWB transmitter part is
highlighted. The normally required area of 0.045 mm^2 of the foundry provided spiral
inductor is reduced to merely 0.003 mm^2 thanks to the active inductor topology.
The measurements were carried out on a chip-on-board package with a 20 GS/s
oscilloscope.

The pulse width control voltage, V_{ctrl}, is swept from 1.4 to 1.8 V, and its effect on
the pulse width is shown in Fig. 4.34. Increasing the voltage increases the switching
speeds of the inverters in the pulse generator and thus makes the pulse width
narrower. The relation between the pulse width and its control voltage is almost
linear. Since the output waveform depends on the start-up and turn-off periods
of the oscillator, the change in the pulse width also affects the amplitude of the
output signal. Figure 4.35 depicts two examples of output waveforms for different
values of V_{ctrl}. The generated minimum pulse width creates an oscillation with
an amplitude of 130 mV. When applying a 200 Mbps data stream, composed by

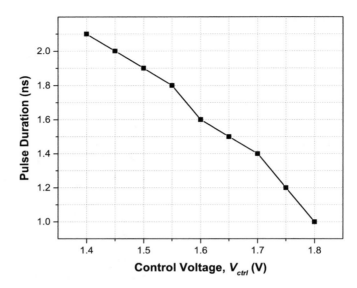

Fig. 4.34 Duration of the pulse with the control voltage, V_{ctrl}

uniformly distributed digital "1"s and "0"s, the measured energy consumption of the circuit is 27.5 and 20 pJ per bit, for the waveforms in Fig. 4.35a and b, respectively.

Figure 4.36 shows the corresponding power spectral densities of the FCC-compliant pulses in Fig. 4.35. As can be seen in Fig. 4.36b, the reduction in amplitude and duration of the pulse results in a broader and lower PSD. Since the center frequency is far enough from the lower limit of the FCC mask, the increase in bandwidth does not cause a violation.

The performance of the designed UWB transmitter is summarized and compared with recently published works in Table 4.3. In addition to the energy consumed per bit, a normalized energy metric, which also covers the area occupation and peak-to-peak output voltage, is introduced. The designed ring oscillator based IR-UWB transmitter shows the smallest area occupation in the literature, while its energy efficiency and pulse amplitude are limited. The main contribution of this work is the minimization of the occupied area by using an active inductor-based local oscillator while maintaining a low power consumption and high data rate communication.

4.4   Summary

This chapter explains how wireless data communication is established for the implanted neural monitoring system. The proposed method offers programmability function for several parameters, such as the active number of electrodes and sampling rate of ADC depending on the situation. A downlink communication

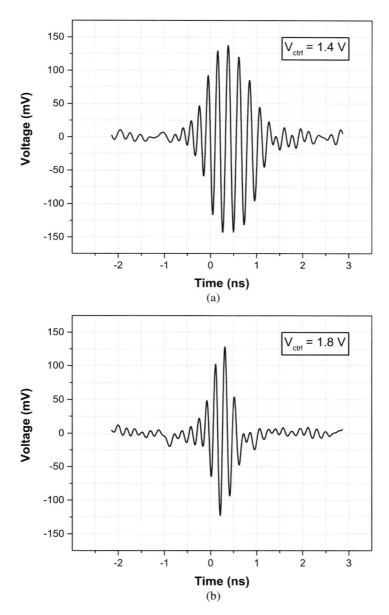

Fig. 4.35 IR-UWB pulse waveforms measured at the transmitter output while (**a**) $V_{ctrl} = 1.4$ V and (**b**) $V_{ctrl} = 1.8$ V

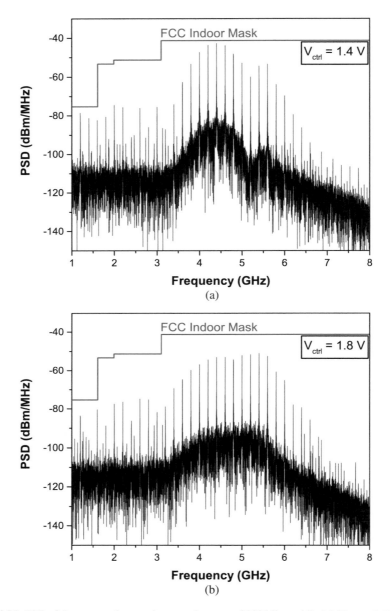

Fig. 4.36 PSD of the proposed transmitter at a data rate of 200 Mbps while (**a**) $V_{ctrl} = 1.4$ V and (**b**) $V_{ctrl} = 1.8$ V

Table 4.3 Comparison of the designed UWB transmitters with similar works in the literature

	Tech. (nm)	Modulation	Method	Frequency band (GHz)	Data rate (Mbps)	Pulse amplitude (mV)	Energy per bit (pJ/b)	Area (core) (mm^2)	Normalized energy (mm^2 · pJ)/(b · V)
MWCL'18 [30]	65	OOK	Edge combining	3.1–4.1	200	110	4.32	0.065	1.27
TCAS-I'18 [36]	130	OOK	Direct, double PLL	3.5–4.5	1000	50	5	0.04	2
JSSC'17 [37]	28	OOK	LO-based	3.5–4.5	27.24	175	14	0.095	3.8
TMTT'17 [38]	180	BPSK-PAM	Edge combining	3.5–6.5	250	500	86	0.22	18.92
JSSC'16 [39]	65	QPSK	LO-based	6.25–8.5	1000	175	221	0.7	442
TBCAS'16 [22]	180	OOK	Edge combining	3.1–7	500	75	7	0.01	0.47
TBCAS'15 [40]	90	OOK	Edge combining	3.1–5	67	255	30	0.061	3.58
TCAS-II'15 [41]	65	PPM + DB-BPSK	Edge combining	3.1–4.8	200	250	30	0.182	10.92
This work[a]	180	OOK	LO-based	3.1–6	250	55.5	45.2	0.0018	0.74
This work[b]	180	OOK	LO-based	3.1–6	200	130	20	0.021	1.61

[a]IR-UWB transmitter based on ring oscillator
[b]IR-UWB transmitter based on active inductor

using low power pulse position modulation is investigated to program the remotely powered implant. Two alternative methods, namely narrowband and ultra-wideband transmitters, are designed to perform transmitting the amplified, quantized, and analyzed neural activities to the external unit. The method for increasing the data rate of the narrowband transmitter is proposed. Additionally, the high area consumption of spiral inductors in oscillators is eliminated by introduced active inductor usage in local oscillators. Finally, the presented circuits are fabricated, and their characterizations are validated with measurement results.

References

1. Schwarz DA, Lebedev MA, Hanson TL et al (2014) Chronic, wireless recordings of large-scale brain activity in freely moving rhesus monkeys. Nat Methods 11:670–676
2. Wang G, Liu W, Sivaprakasam M, Kendir GA (2005) Design and analysis of an adaptive transcutaneous power telemetry for biomedical implants. IEEE Trans Circuits Syst Regul Pap 52:2109–2117
3. Tang Z, Smith B, Schild JH, Peckham PH (1995) Data transmission from an implantable biotelemeter by load-shift keying using circuit configuration modulator. IEEE Trans Biomed Eng 42:524–528
4. Bohorquez JL, Chandrakasan AP, Dawson JL (2009) A 350μW CMOS MSK transmitter and 400μW OOK super-regenerative receiver for medical implant communications. IEEE J Solid State Circuits 44:1248–1259
5. Chae MS, Yang Z, Yuce MR et al (2009) A 128-channel 6 mW wireless neural recording IC with spike feature extraction and UWB transmitter. IEEE Trans Neural Syst Rehabil Eng 17:312–321
6. Ballini M, Müller J, Livi P et al (2014) A 1024-channel CMOS microelectrode array with 26,400 electrodes for recording and stimulation of electrogenic cells in vitro. IEEE J Solid State Circuits 49:2705–2719
7. Liu W, Vichienchom K, Clements M et al (2000) A neuro-stimulus chip with telemetry unit for retinal prosthetic device. IEEE J. Solid State Circuits 35:1487–1497
8. Lee H, Kim J, Ha D et al (2015) Differentiating ASK demodulator for contactless smart cards supporting VHBR. IEEE Trans Circuits Syst Express Briefs 62:641–645
9. Ghovanloo M, Najafi K (2004) A wideband frequency-shift keying wireless link for inductively powered biomedical implants. IEEE Trans Circuits Syst Regul Pap 51:2374–2383
10. Hwang Y, Hwang B, Lin H, Chen J (2013) PLL-based contactless energy transfer analog FSK demodulator using high-efficiency rectifier. IEEE Trans Ind Electron 60:280–290
11. Lee S, Hsieh C, Yang C (2012) Wireless front-end with power management for an implantable cardiac microstimulator. IEEE Trans Biomed Circuits Syst 6:28–38
12. Hu Y, Sawan M (2005) A fully integrated low-power BPSK demodulator for implantable medical devices. IEEE Trans Circuits Syst Regul Pap 52:2552–2562
13. Gong CA, Shiue M, Yao K et al (2008) A truly low-cost high-efficiency ASK demodulator based on self-sampling scheme for bioimplantable applications. IEEE Trans Circuits Syst Regul Pap 55:1464–1477
14. Wang C, Chen C, Kuo R, Shmilovitz D (2010) Self-sampled All-MOS ASK demodulator for lower ISM band applications. IEEE Trans Circuits Syst Express Briefs 57:265–269
15. Kilinc EG, Dehollain C, Maloberti F (2014) A low-power PPM demodulator for remotely powered batteryless implantable devices. In: 2014 IEEE 57th international midwest symposium on circuits and systems (MWSCAS). pp 318–321

16. RamRakhyani AK, Mirabbasi S, Chiao M (2011) Design and optimization of resonance-based efficient wireless power delivery systems for biomedical implants. IEEE Trans Biomed Circuits Syst 5:48–63

17. Muller R, Le H, Li W et al (2015) A minimally invasive 64-channel wireless μECoG implant. IEEE J Solid State Circuits 50:344–359

18. Tan J, Liew W, Heng C, Lian Y (2014) A 2.4 GHz ULP reconfigurable asymmetric transceiver for single-chip wireless neural recording IC. IEEE Trans Biomed Circuits Syst 8:497–509

19. Kassiri H, Bagheri A, Soltani N et al (2016) Battery-less tri-band-radio neuro-monitor and responsive neurostimulator for diagnostics and treatment of neurological disorders. IEEE J Solid State Circuits 51:1274–1289

20. Kiani M, Ghovanloo M (2013) A 20-Mb/s pulse harmonic modulation transceiver for wideband near-field data transmission. IEEE Trans Circuits Syst Express Briefs 60:382–386

21. Bourdel S, Bachelet Y, Gaubert J et al (2010) A 9-pJ/pulse 1.42-Vpp OOK CMOS UWB pulse generator for the 3.1–10.6-GHz FCC band. IEEE Trans Microwave Theory Tech 58:65–73

22. Mirbozorgi SA, Bahrami H, Sawan M et al (2016) A single-chip full-duplex high speed transceiver for multi-site stimulating and recording neural implants. IEEE Trans Biomed Circuits Syst 10:643–653

23. Mercier PP, Daly DC, Chandrakasan AP (2009) An energy-efficient all-digital UWB transmitter employing dual capacitively-coupled pulse-shaping drivers. IEEE J Solid State Circuits 44:1679–1688

24. Hajimiri A, Lee TH (1999) Design issues in CMOS differential LC oscillators. IEEE J Solid-State Circuits 34:717–724

25. Ham D, Hajimiri A (2001) Concepts and methods in optimization of integrated LC VCOs. IEEE J Solid State Circuits 36:896–909

26. Jung J, Zhu S, Liu P et al (2010) 22-pJ/bit energy-efficient 2.4-GHz implantable OOK transmitter for wireless biotelemetry systems: in vitro experiments using rat skin-mimic. IEEE Trans Microwave Theory Tech 58:4102–4111

27. Smith G (1971) Radiation efficiency of electrically small multiturn loop antennas used in upper atmosphere propagation experiments. In: 1971 Antennas and Propagation Society International Symposium, pp 113–116

28. Smith G (1972) Radiation efficiency of electrically small multiturn loop antennas. IEEE Trans Antennas Propag 20:656–657

29. Federal Communications Commission (2002) First report and order regarding ultra-wideband transmission systems

30. Lin Y, Park S, Chen X et al (2018) 4.32-pJ/b, overlap-free, feedforward edge-combiner-based ultra-wideband transmitter for high-channel-count neural recording. IEEE Microwave Wireless Compon Lett 28:52–54

31. Mir-Moghtadaei SV, Fotowat-Ahmady A, Nezhad AZ, Serdijn WA (2014) A 90 nm-CMOS IR-UWB BPSK transmitter with spectrum tunability to improve peaceful UWB-narrowband coexistence. IEEE Trans Circuits Syst Regul Pap 61:1836–1848

32. Kim N, Rabaey JM (2016) A high data-rate energy-efficient triple-channel UWB-based cognitive radio. IEEE J Solid State Circuits 51:809–820

33. Bahrami H, Mirbozorgi SA, Ameli R et al (2016) Flexible, polarization-diverse UWB antennas for implantable neural recording systems. IEEE Trans Biomed Circuits Syst 10:38–48

34. Craninckx J, Steyaert M (1995) Low-noise voltage-controlled oscillators using enhanced LC-tanks. IEEE Trans Circuits Syst II Analog Digit Signal Process 42:794–804

35. Lu L-H, Hsieh H-H, Liao Y-T (2006) A wide tuning-range CMOS VCO with a differential tunable active inductor. IEEE Trans Microwave Theory Tech 54:3462–3468

36. Crepaldi M, Angotzi GN, Maviglia A et al (2018) A 5 pJ/pulse at 1-Gpps pulsed transmitter based on asynchronous logic master–slave PLL synthesis. IEEE Trans Circuits Syst Regul Pap 65:1096–1109

37. Streel G de, Stas F, Gurné T et al (2017) SleepTalker: a ULV 802.15.4a IR-UWB transmitter SoC in 28-nm FDSOI achieving 14 pJ/b at 27 Mb/s with channel selection based on adaptive FBB and digitally programmable pulse shaping. IEEE J Solid State Circuits 52:1163–1177

38. Gunturi P, Emanetoglu NW, Kotecki DE (2017) A 250-Mb/s data rate IR-UWB transmitter using current-reused technique. IEEE Trans Microwave Theory Tech 65:4255–4265
39. Ko J, Gharpurey R (2016) A pulsed UWB transceiver in 65 nm CMOS with four-element beamforming for 1 Gbps meter-range WPAN applications. IEEE J Solid State Circuits 51:1177–1187
40. Ebrazeh A, Mohseni P (2015) 30 pJ/b, 67 Mbps, centimeter-to-meter range data telemetry with an IR-UWB wireless link. IEEE Trans Biomed Circuits Syst 9:362–369
41. Na K, Jang H, Ma H et al (2015) A 200-Mb/s data rate 3.1–4.8-GHz IR-UWB all-digital pulse generator with DB-BPSK modulation. IEEE Trans Circuits Syst Express Briefs 62:1184–1188

Chapter 5
Experimental Validations

5.1 Introduction

In Chaps. 3 and 4, the circuits for remote powering of the implanted system and wireless communication methods are presented, respectively. In order to verify the performances of the individual blocks and the integration of the system, the proposed circuits are fabricated using UMC 180 nm MM/RF technology. Figure 5.1 represents the final version of the fabricated chips which includes all the blocks explained in previous chapters. The dimensions of the die are 15 mm × 15 mm.

The measurements for the characterization of each block are carried out with chip-on-board packages. The dies are bound to the PCB with the help of a conductive adhesive. Gold wire bonding with 20 µm thickness is used to connect the pads on the integrated circuit to the PCB. The wedge-wedge type connections were made in the cleanroom facility at EPFL. Figure 5.2 shows the integration of the die and the PCB.

The die and wire bondings are encapsulated with non-conductive glob top to protect from the mechanical strain, scratching, dust, and dirt. Glob top provides electrical insulation, mechanical support, and a conservative environment from light and moisture for better characterization.

5.2 Biomedical Packaging

There are several regulations and standards to be ensured prior to the marketing of the implantable device. In the United States (US), every marketed medical device has to be approved by the Food and Drug Administration (FDA), whereas they must carry Conformité Européenne (CE) mark in the European Union (EU). Although there are some differences in requirements and procedures, the primary goal of both

© Springer Nature Switzerland AG 2020   91
K. Türe et al., *Wireless Power Transfer and Data Communication for Intracranial Neural Recording Applications*, Analog Circuits and Signal Processing,
https://doi.org/10.1007/978-3-030-40826-8_5

Fig. 5.1 Circuit blocks on the latest chip: (a) narrowband transmitter, (b) ultra-wideband transmitter based on active inductor, (c) power feedback generator, (d) rise detector, (e) automatic resonance tuning, and (f) active half-wave rectifier, low drop-out regulator, downlink demodulator, and narrowband transmitter

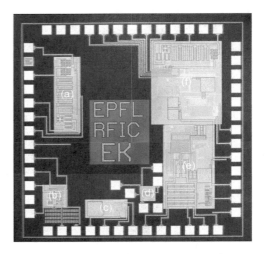

Fig. 5.2 Wire bonded chip-on-board

of them is to ensure that the medical devices are safe for the body and effective regarding the operation. There are mainly three probable of biocompatibility issues. Firstly, the implant may contain toxic substances that may be hazardous to the body and may create inflammation. Secondly, the body may consider the implant as a foreigner, and the immune system starts to fight against the device even if there is nothing dangerous for the body. Lastly, the implanted device may stop working because of being inversely affected by the tissue. Even worse case, the affected device may start malfunctioning and begin to harm the body. For every example, the implanted device should be removed right away. In order to prevent those unwanted probabilities, biocompatibility test is a must in both US and EU.

The International Organization for Standardization (ISO) is a worldwide federation of national standards bodies. In 2018, ISO released the document with the identification of ISO 10993-1:2018 which is the 5th edition of biocompatibility standard for evaluation of medical devices. The standard provides a general categorization of medical devices based on the nature and duration of their contact with the body. Moreover, the required evaluations are presented for the intended use of the medical device. Figure 5.3 summarizes the categorizations regarding contact type and contact duration as well as the necessary tests to be pass in order to be considered as biocompatible.

The implantable medical devices are subcategorized concerning the contact point as tissue/bone and blood. Furthermore, regarding the contact duration of the medical device, there are three groups as follows:

- limited duration (less than 24 h),
- prolonged duration (between 24 h and 30 days),
- long-term duration (more than 30 days).

A presurgical analysis of epilepsy disease can take several days and even several weeks. Therefore, according to ISO 10993-1:2018, the implantable system for presurgical monitoring of neural activity should be considered as prolonged contact and should satisfy tests for cytotoxicity, sensitization, irritation or intracutaneous reactivity, material mediated pyrogenicity, acute systemic toxicity, subacute toxicity, implantation effects, and genotoxicity to be considered as biocompatible.

5.3 Animal Experiments

The measurements for each block presented in previous chapters are conducted in a laboratory environment with ideal conditions. However, both FDA and CE impose validation of safety and success with clinical studies before the marketing of the implantable device. To show the functionality in the targeted environment, different animal experiments were conducted for *in vivo* characterization of remote powering and wireless data communication.

Animal experiments were approved by the Animal Research Ethics Committee of the Canton of Bern, Switzerland. The tests were conducted under the supervision of Prof. Hans Rudolf Widmer and with the assistance of Dr. Stefano Di Santo, Dr. Stefanie Seiler, and Dr. Thuy Anh Khoa Nguyen in the Department of Neurology at Bern University Hospital.

The experiments were planned in three following steps

1. characterization of remote powering with 4-coils, active half-wave rectifier, and low drop-out regulator,
2. characterization of wireless data communication with the narrowband transmitter,
3. wireless data transmission of acquired neural activity.

A printed circuit board is designed to host the integrated circuits and external components. The dimensions of the board are 20 mm × 20 mm. Figure 5.4 shows

Medical device categorization by / **Nature of body contact** — **Biological Effect**

Contact duration: A - limited (≤24 h); B - prolonged (>24 h to 30 d); C - permanent (>30 d)

Category	Contact	Contact duration	Cytotoxicity	Sensitization	Irritation or intracutaneous reactivity	Material mediated pyrogenicity	Acute systemic toxicity	Subacute toxicity	Subchronic toxicity	Chronic toxicity	Implantation effects	Hemocompatibility	Genotoxicity	Carcinogenicity
Surface device	Skin	A	X	X	X									
		B	X	X	X									
		C	X	X	X									
	Mucosal membrane	A	X	X	X									
		B	X	X	X			X	X					
		C	X	X	X			X		X			X	
	Breached or compromised surface	A	X	X	X	X	X	X						
		B	X	X	X	X	X	X	X		X			
		C	X	X	X	X	X	X		X	X		X	X
External communicating device	Blood path, indirect	A	X	X	X	X	X	X				X		
		B	X	X	X	X	X	X	X			X		
		C	X	X	X	X	X	X		X		X	X	
	Tissue/bone/ dentin	A	X	X	X	X	X	X						
		B	X	X	X	X	X	X	X		X		X	
		C	X	X	X	X	X	X		X	X		X	X
	Circulating blood	A	X	X	X	X	X	X				X	X	
		B	X	X	X	X	X	X	X		X	X	X	
		C	X	X	X	X	X	X		X	X	X	X	X
Implant device	Tissue/bone	A	X	X	X	X	X	X						
		B	X	X	X	X	X	X	X		X		X	
		C			X					X	X		X	X
	Blood	A	X	X	X	X	X	X			X	X	X	
		B	X	X	X	X	X	X	X		X	X	X	
		C			X					X	X	X	X	X

Fig. 5.3 ISO 10993-1 biocompatibility testing selection criteria

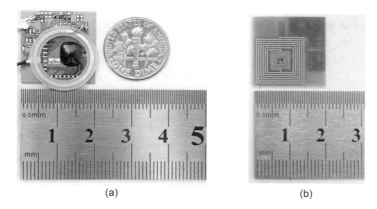

(a)   (b)

Fig. 5.4 (**a**) Top view and (**b**) bottom view of the boards for animal experiments

the fabricated and the assembled board. The external capacitors are connected to the output of the rectifier and regulator blocks. As it presented in Chap. 3, the narrowband transmitter has a variable capacitor and a 3-bit capacitor bank for fine-tuning of the carrier frequency. In addition to the external capacitors, there are $0\,\Omega$ resistances connected between the control bits of the capacitor bank and either regulated output voltage or ground to activate or deactivate control bits, respectively.

The assembled board has to be packaged with a biocompatible material. Titanium is the most commonly used packaging material, especially for cardiac pacemakers. However, its stiffness creates a drawback for the intended neural monitoring application. The microelectrode array is planned to be placed on the cortex and flexibility of the packaging is essential. Moreover, the titanium behaves like a Faraday cage and does not allow the passage of the radio frequency (RF) signals. This situation disables remote powering and wireless data communication functions of the implanted device. One solution to overcome this issue is to make an opening for the RF signals and cover this portion of the implanted device with glass. It is a promising solution; however, it requires a complex process and expertise on the packaging.

As an alternative to the titanium and any other metal case, polymer-based materials shows a comparable performance for encapsulation and packaging. Polymers present fast implantation, long-term stability, higher flexibility compared to metal and ceramic, good electrical insulation while not blocking the electromagnetic radiations [1]. The most commonly used polymers are polyimide, polydimethylsiloxane (PDMS), Parylene-C, and epoxy.

The conducted animal experiments in the scope of this book focus on the functionality of the implantable system without underestimating the biocompatibility issues. Therefore, easy-to-use, off-the-shelve biocompatible encapsulation materials are investigated. The search for the packaging material is ended up with silicone adhesive, Silastic Medical Adhesive Silicone, Type A by Dow Corning Corporation.

Fig. 5.5 Implantable board
with biomedical packaging

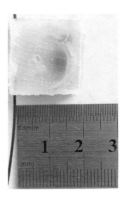

The selected silicone offers to cure at room temperature and does not require an additional solvent. Moreover, the manufacturer claims that the material has passed screening tests that apply to products intended for implantation for up to 29 days. Accordingly, the designed board for *in vivo* characterization is encapsulated with the selected material. Figure 5.5 shows the circuits with biomedical packaging. Dimensions of the electronics became 21 mm × 21 mm with 3 mm thickness after packaging.

5.3.1 *In Vivo Characterization of Remote Powering*

The proposed implantable system for intracranial neural monitoring targets human applications. However, a rat is chosen for the animal experiments for the first phase of the validations, and the available space is limited in the skull of a rat. Accordingly, the remote powering and wireless data communication circuits are tested by implanting them into its ventral region. The belly part of the rat provides enough space for the circuits and enables efficient remote powering by minimizing the distance between the implant and external coils, which are planned to be placed underneath the rat.

During the implantation, an opening about 30 mm is created on the skin of the anesthetized rat. The fat tissue between the muscles and the skin is removed, and the device is placed between the skin and muscle. The orientation of the implant is selected such that the two powering coils were closer to the skin to decrease the distance to the external ones. After the placement of the implant, the opening was closed with three stitches. Figure 5.6 shows the rat with the implanted device in its ventral region.

The rat is placed on top of the external powering coils as it is presented in Fig. 5.7. A signal generator drives the inner coil on the external board at 8 MHz frequency. An external resistor connected to the LDO regulator output in order to mimic the predicted system load of 10 mW. Figure 5.8 illustrates the rectified and regulated

Fig. 5.6 The rat with implanted device into its ventral region

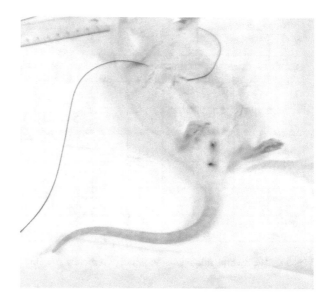

Fig. 5.7 The rat with implanted device placed on powering coils

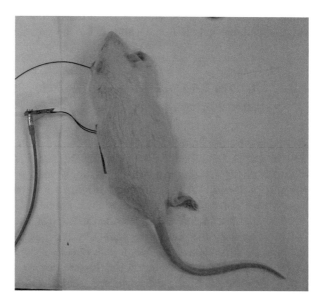

output voltages. The laboratory measurements show the efficiencies for the 4-coil inductive link, active half-wave rectifier, and LDO regulator are 55%, 82%, and 78%, respectively. The overall power delivering efficiency of the remote powering chain starting from the external base station to 10 mW regulator load is expected as 36%. However, the power transfer efficiency during the *in vivo* characterization

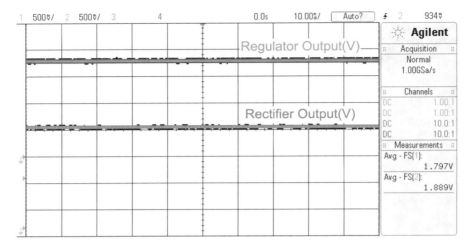

Fig. 5.8 The rectifier and regulator output of remotely powered system

is measured as 30%. The power transfer efficiency goes down from the expected value because of the tissue of the rats, packaging, and imprecise alignment between external and internal coils, and a shift in the resonance frequency of internal coils.

5.3.2   *In Vivo Characterization of Wireless Data Communication*

The procedure for the implantation of the board with the narrowband transmitter was the same as the remote powering implant. Again, two cables were coming out of the rat's belly. One cable provides the supply voltage to the implant, and the other one carries the data for the transmitter. Figure 5.9 shows the rat with the implanted narrowband transmitter with the monopole antenna (TI.10.0112) which is placed 30 cm away from the rat to measure the radiation spectrum.

The laboratory measurements show that the designed transmitter is capable of delivering information at 12 Mbps data rate. A square wave at 6-MHz frequency is applied to the data input of the transmitter to test its functionality at its limit. The antenna in Fig. 5.9 is directly connected to a spectrum analyzer and Fig. 5.10 displays the measured spectrum of the transmitter. The carrier frequency is measured as 429 MHz with −59.4 dBm power. The applied data generated two peaks placed 12 MHz away from the carrier frequency as expected.

An external receiver circuit is designed using off-the-shelve components to decode the captured OOK modulated signal by the antenna. Figure 5.11 shows the receiver topology with the models and brands of the components. The received signal is amplified by the low noise amplifier (LNA) and down-converted with

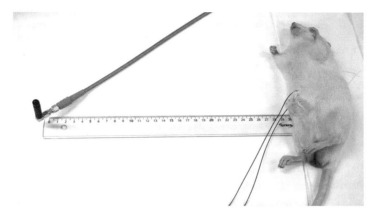

Fig. 5.9 The rat with implanted narrowband transmitter and receiver antenna

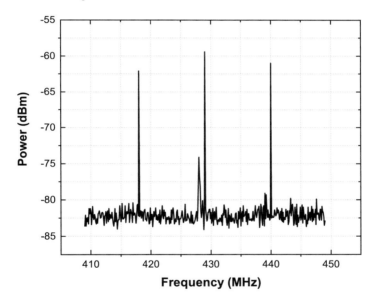

Fig. 5.10 The measured spectrum of the implanted narrowband transmitter

the mixer. The down-converted frequency is determined by a voltage controlled oscillator (VCO). The logarithmic amplifier converts the input power to an output voltage. Therefore, the presence and absence of an incoming signal generate different voltage values at the output of the logarithmic amplifier. At the final stage of the receiver chain, a comparator distinguishes the different voltages for the modulated signal and provides rail-to-rail outputs for "0" and "1" bits.

For testing the functionality of transmitter, a uniformly distributed data as "...010101" was applied to the implanted transmitter and a custom-designed board

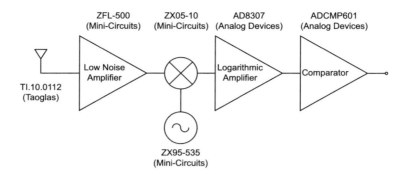

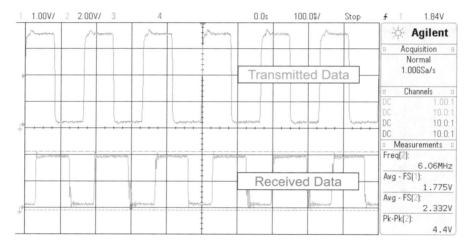

Fig. 5.11 Block diagram of receiver used in *in vivo* characterization

Fig. 5.12 The measurement of received data at distance of 30 cm

was used for receiving the signal. The receiver board was placed 30 cm away from the rat, and a communication link with 12 Mbps data rate was established, as presented in Chap. 4. Figure 5.12 depicts the oscilloscope screen that shows the data applied to the transmitter and the received signal with the external base station.

Table 5.1 summarizes the measurement results for remote powering and wireless data communication with a narrowband transmitter during *in vivo* characterizations.

5.3.3   System Level In Vivo Characterization

For the verification of the proposed system, the last experiment is planned to measure the neural activity of a rat using available components on the market and to transmit the data via the designed transmitter. The purpose of this experiment is to

Table 5.1 Summary of *in vivo* characterized remote powering and wireless data communication blocks

Remote powering	
Powering frequency (MHz)	8
Powering distance (mm)	10
Powering efficiency (%)	30
Delivered power (mW)	10
Wireless data communication	
Modulation type	OOK
Carrier frequency (MHz)	429
Data rate (Mbps)	12
Communication distance (cm)	30
Power consumption (μW)	544
Energy/bit (pJ/b)	44

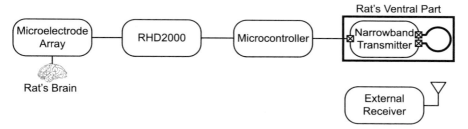

Fig. 5.13 Block diagram of wireless transmission of neural measurement

show the capability of the proposed neural monitoring system. The block diagram of the system level experiment is presented in Fig. 5.13.

A PCB containing the narrowband transmitter block is designed, packaged with a biocompatible material, and placed in the belly region of the rat as previously performed. Two cables were coming out of the body; the first cable is for supplying 1.8 V for the implanted transmitter and the second cable is for delivering the digitalized measurement result to the data input of the wireless data communication.

In addition to the implantation of the board into the ventral region, another surgery is needed for the placement of the surface electrode array. The rat with a shaved head was fixed on a stereotaxic frame for the craniotomy. A midline scalp incision was applied to reach the skull. An opening created by the help of a drill and the dura is removed to reach the cortex. Two screws were placed on the skull as a reference point for the electrodes. The surface electrode array is put on the cortex and ground wires are connected to the screws as presented in Fig. 5.14.

A microelectrode array with 32 channels from NeuroNexus (E32-600-10T-100) is chosen for capturing the neural activities. The substrate material of the electrode is polyimide and platinum is used as electrode site material. The distribution of the electrode array and its dimensions are depicted in Fig. 5.15.

Fig. 5.14 Placed MEA on
the cortex of the rat and
reference screws

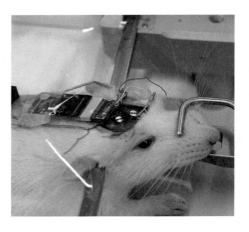

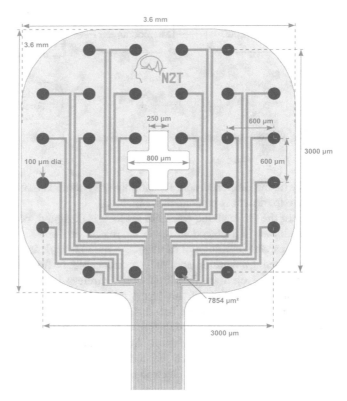

Fig. 5.15 Illustration of the surface MEA with its dimensions. Adapted from [2]

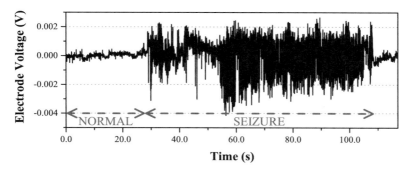

Fig. 5.16 Recorded and transmitted brain waveform of a rat during a seizure

The electrodes are connected to an RHD2000 series digital electrophysiology interface chip from Intan Technologies. The lower and upper cutoff frequencies are selected as 1 and 500 Hz, respectively. The recording chip amplifies the neural signal by 45.6 dB and converts to the digital domain with 16-bit ADC. The serial peripheral interface (SPI) output of the neural recording chip is converted to a bit-stream by a micro-controller unit. The measurement results are transmitted by the designed narrowband transmitter and decoded using the aforementioned receiver. Figure 5.16 represents the recorded and transmitted brain waveform of a rat during a seizure.

5.4 Summary

In this chapter, the fabricated integrated circuit for remote powering and wire-less data communication blocks of the implantable neural monitoring system is presented. The importance of biomedical packaging is discussed, and an easy-to-apply solution for the proposed system is introduced. The functionality of the remote powering system and the narrowband transmitter is verified *in vivo* with implantation into the rat. A receiver circuit for wireless data communication is designed with discrete components. Finally, the proposed system is validated with off-the-shelve components and *in vivo* measurements of neural recordings.

References

1. Hassler C, Boretius T, Stieglitz T (2011) Polymers for neural implants. J Polym Sci Part B: Polym Phys 49:18–33
2. E32-600-10-100 NeuroNexus. In: NeuroNexus. http://neuronexus.com/electrode-array/e32-600-10-100/. Accessed 11 Dec 2018

Chapter 6
Conclusion

6.1 Concluding Remarks

In the scope of this book, an implantable and entirely wireless intracranial neural monitoring system for presurgical analysis of epilepsy is proposed. The primary goal of the method is to eliminate the wires coming out from the electrodes in a conventional intracranial EGG. Additionally, the suggested system will improve the patient's comfort by excluding the situation of being connected to an external measurement unit.

Firstly, the system-level approach to the fully implanted intracranial monitoring of epilepsy is developed. The challenges in the implantable system regarding size, temperature elevation, power source, communication speed, and packaging are discussed in detail. The solutions to overcome the implantation difficulties and circuit blocks needed to develop the proposed system are analyzed. On the circuit level, the work presented in the book focuses on remote powering of the implant, wireless data communication between the implant and external unit, and *in vivo* characterization of the designed circuits with biomedical packaging.

For the energy source of the implanted system, the battery provides comfort and freedom in movements to the patient. Collection and wireless transmission of the excessive amount of data restrict the operation time of the battery. The remote powering is presented for recharging energy storage element. Regarding the transmission distance and the required power level, inductive coupling is selected as the best candidate. A 4-coil inductive link is used for remote powering since it shows better power transfer efficiency and less sensitivity to load variations compared to the conventional 2-coil structure. The magnetically induced AC voltage is converted to stable and ripple-free supply voltage by an active half-wave rectifier and a low drop-out voltage regulator. The efficiency of the power conversion is analyzed, and an improving technique is introduced to limit temperature elevation due to

© Springer Nature Switzerland AG 2020   105
K. Türe et al., *Wireless Power Transfer and Data Communication for Intracranial Neural Recording Applications*, Analog Circuits and Signal Processing,
https://doi.org/10.1007/978-3-030-40826-8_6

the dissipated power. Moreover, an automatic tuning system is developed to match the LC resonance frequency of the implanted coil with remote powering frequency. Furthermore, the power feedback generation structure is designed to make sure that the delivered power is high enough for the intended operation.

In addition to supplying power to the implantable device, establishing the data communication between the device and the external unit is an important challenge. The controlled changes in the remote powering signal can be used as a mean of data transfer from the external station to the implant. The introduced pulse position modulation enables the transmission of data on the inductive link while maintaining reliable received power. Two alternative methods, namely narrowband and ultra-wideband transmitters, are designed to transmit the amplified, quantized, and analyzed neural activities to the external unit. A new approach is developed for increasing the data rate in on-off keying modulated narrowband signal while the consumed power is minimally affected. Additionally, the use of an active inductor in impulse radio ultra-wideband (IR-UWB) communication is proposed to reduce the area occupied by the conventional spiral inductor.

The proposed remote powering and wireless data communication circuits are designed, fabricated, and tested in the laboratory environment. The *in vivo* characterization of the system requires to meet regulations and standards for packaging and approval by the animal research ethics committee. The circuits are encapsulated with silicone-based commercially available adhesive material, which satisfies the safety criteria. Remote powering and narrowband transmitter integrated chips are tested in the case of a rat. Moreover, the system-level validation is achieved by a seizure detection with discrete components and transmission by the implanted transmitter.

The main innovations presented in this book can be summarized as follows.

- A low power downlink communication method based on pulse position modulation (PPM) is developed for remotely powered systems. PMM modulation provides stable powering performance while establishing a downlink data communication. Besides, the data transfer is not affected by the misalignments occurring between the powering coils.
- A technique is developed for optimization of the rise and fall times of a cross-coupled pair LC voltage controlled oscillator for high communication data rates. The proposed optimization procedure improves the maximum data rate for OOK modulation by a factor of 87% while the DC power consumption increases by 6.7%.
- An IR-UWB transmitter based on an active inductor is designed, fabricated, and characterized. The main contribution consists of replacing the conventional spiral inductor with an active one. The area is drastically decreased while keeping low power consumption and a high data rate.

6.2   Future Work

The proposed neural monitoring system offers higher temporal and spatial resolutions than traditional EEG and eliminates possible complications created by the available intracranial EEG. The remote powering and wireless data communication blocks of the implantable system are characterized *in vivo*. However, further work, improvements, and experiments need to be carried out to establish an operating device for epilepsy patients. The recommended future studies on the presented system are discussed as follows.

- The automatic resonance tuning algorithm in this study is performed to tune only one LC resonator in 2-coil inductive link. A possible future step of this work is to extend the tuning mechanism to tune two LC resonators to make the benefit of using a 4-coil link as presented in this work.
- For the energy source of the implanted system, a hybrid solution composed of a rechargeable battery and remote powering operation is proposed. Remote powering circuits are designed and verified thanks to a commercial battery charger. Future work should focus on the CMOS integration of remote powering block and battery charger block.
- Although remote powering and data transmission circuits are characterized in animal experiments, further experimental investigations are needed at the system level to detect epilepsy in multiple rodents such as rat or mice.
- Extraction of reliable medical information with the proposed system and integration with a neurostimulator circuit can enable a closed-loop system to decrease the effects of epilepsy seizures drastically.

Index

© Springer Nature Switzerland AG 2020
K. Türe et al., *Wireless Power Transfer and Data Communication for Intracranial Neural Recording Applications*, Analog Circuits and Signal Processing,
https://doi.org/10.1007/978-3-030-40826-8

Printed in the United States
By Bookmasters